Social-Emotional Skills, Resilience, and Behavioral Problems in Children

Social-Emotional Skills, Resilience, and Behavioral Problems in Children

Guest Editor

Stefano Eleuteri

MDPI

Basel • Beijing • Wuhan • Barcelona • Belgrade • Novi Sad • Cluj • Manchester

Guest Editor
Stefano Eleuteri
Department of Human and
Social Sciences
Universitas Mercatorum
Rome
Italy

Editorial Office
MDPI AG
Grosspeteranlage 5
4052 Basel, Switzerland

This is a reprint of the Special Issue, published open access by the journal *Children* (ISSN 2227-9067), freely accessible at: https://www.mdpi.com/journal/children/special_issues/1124KR9023.

For citation purposes, cite each article independently as indicated on the article page online and as indicated below:

Lastname, A.A.; Lastname, B.B. Article Title. *Journal Name* **Year**, *Volume Number*, Page Range.

ISBN 978-3-7258-7827-7 (Hbk)
ISBN 978-3-7258-7828-4 (PDF)
https://doi.org/10.3390/books978-3-7258-7828-4

Contents

About the Editor

Stefano Eleuteri

Stefano Eleuteri is an assistant professor of clinical psychology at Mercatorum University, Rome. He is a health psychologist, psychotherapist and sexologist with a PhD in social developmental psychology and educational research. He is a member of the executive committee of the World Association for Sexual Health and of the European Federation of Sexology. Since 2010, he has worked as a psychologist, psychotherapist, sexologist and as a private consultant in the field of sexual education, also organizing events about sexual health.

He is dedicated to understanding changes in the developmental stages connected to environmental changes. His main research interests are connected with sexual health and sociocultural aspects, especially the modification of sexuality connected with new media, but also the psychological changes connected to sociocultural aspects. He has spoken at the most important local and international congresses in the sexological sciences field, presenting works in the clinical sexology and sexual health. He has published articles on science and education in the field of psychology, and also in indexed journals.

Preface

Childhood and adolescence are crucial periods for the development of social–emotional skills. Optimal development in these areas during early life is associated with enhanced social and cognitive abilities, better academic performance, motivation, and social relationships, a reduced risk of behavioral problems (internalizing or externalizing) and social exclusion, and higher rates of overall well-being. This Reprint aims to explore the role of protective factors like socio-emotional learning and resilience in promoting prosocial behavior in children. I believe that focusing on the emotional aspect of development is a key challenge for schools in the new millennium. In this Reprint, you will find articles on childhood and adolescents, focused on the active promotion of these skills through the implementation of educational programs. The articles provide interesting insights and important contributions to children's development in the social–emotional domain. Authors explored what schools can do to contribute to the social and emotional education of children and adolescents and engaged with how schools can develop these aspects. Different evolutionary problems are described: preschool exclusionary discipline, weight stigma, gender and sexual minority stress, etc. Specific pathologies are considered, like autism spectrum disorder and anxiety disorders. For example, sensory modulation using both self-reports (to reflect implications in real life from patients' own voices) along with objective measures are used to explain daily behaviors by underlying physiological mechanisms in anxiety disorders. Authors suggest that it may focus intervention towards better health, function, and development. Using a positive psychology approach, not just negative issues are taken into account: creativity, socioemotional strengths and sexuality education are offered as strategies to improve psychological wellbeing in children and adolescents. An article focuses on the development and validation of the Children's Emotions Database (CED), a tool to evaluate preschoolers' basic and complex facial expressions.

It also expanded how early childhood education, in particular the interaction between teachers and children and peer interactions, contributes to children and adolescents developing social–emotional skills. One review analyzes also the possibility of extending sexual education to sexual and gender minorities. The reprint is addressed to scholars, psychologists, educators working with children or adolescents. It offers, in fact, not just new research data, but also clinical and educational insights.

Stefano Eleuteri
Guest Editor

children

MDPI

Systematic Review

A Systematic Review of Treatment for Children with Autism Spectrum Disorder: The Sensory Processing and Sensory Integration Approach

Jonathan Camino-Alarcón [1], Maria Auxiliadora Robles-Bello [1,*], Nieves Valencia-Naranjo [1] and Aziz Sarhani-Robles [2]

[1] Department of Psychology, University of Jaen, 23071 Jaen, Spain; jvca0001@red.ujaen.es (J.C.-A.); nnaranjo@ujaen.es (N.V.-N.)

[2] Medicine Faculty, University of Granada, 18016 Granada, Spain; asarhanir@correo.ugr.es

* Correspondence: marobles@ujaen.es

Abstract: Background/Objectives: The prevalence of the diagnosis of autism spectrum disorder (ASD) has been increasing globally, necessitating updates to the Diagnostic and Statistical Manual of Mental Disorders with respect to ASD diagnosis. It is now recognised that ASD is related to sensory processing disorder, and sensory integration is considered a suitable intervention for treating children diagnosed with ASD. Methods: This paper provides a systematic review on a timeline from 2013 to 2023, based on the PRISMA model. Evidence was sought in the academic search engines Pubmed, Scielo, Eric, Dialnet, Springer, Base Search and Google Scholar, which produced 16 articles according to the inclusion criteria. Results: According to the results of this review, intervention with sensory integration in infants with ASD meets the criteria to be considered an evidence-based practice. The studies reviewed focused mainly on clinical settings and, therefore, we highlight the urgent need for further research to evaluate the effectiveness of sensory integration interventions in naturalistic settings such as homes and schools. Conclusions: This will help to obtain more representative data on how these interventions affect the daily lives of children with ASD.

Keywords: sensory integration; assessment; treatment; autism spectrum disorder; systematic review

1. Introduction

Today, the term autism spectrum disorder (ASD) is used to describe a clinically heterogeneous group of neurodevelopmental disorders that share core behavioural features that affect social communication and include stereotypic, restrictive and repetitive behaviour patterns and interests. This term refers to several conditions as idiopathic forms, including autism, Asperger's syndrome, pervasive developmental disorders not otherwise specified, childhood disintegrative disorder and Rett syndrome. With the establishment of the Diagnostic and Statistical Manual of Mental Disorders (DSM-5), these previously separate diagnoses have been grouped under a unifying umbrella called ASD. ASD is also associated with many comorbidities, which can be physiological and psychological [1]. These changes were introduced to create a simpler and clearer system for identifying and diagnosing ASD, based on the number of specific symptoms or deficits needed to meet the diagnostic criteria [2] (Peters and Matson, 2020).

However, ASD research is heavily biased towards high-income Western countries. In most low- and middle-income countries, where the majority of people with ASD worldwide live, culturally appropriate ASD screening and diagnostic tools are lacking. Although the domains that define ASD are found at a global level, subtle differences in specific manifestations of symptoms, at the subdomain or behavioural level, may exist in different ethnic groups. Therefore, cultural differences in the expression of ASD should be taken into account, e.g., in Chinese culture, eye contact with authority is traditionally considered shameful, whereas in some African or Latin cultures, it is considered disrespectful. In these cultures, children who do not make eye contact with parents, teachers, or physicians may be perceived as compliant rather than exhibiting atypical autistic behaviours, so the eye contact abnormalities described in the DSM-5 may be more subtle and difficult to identify. On the other hand, in India, there is a common local cultural belief that boys start speaking later than girls, while in China, a language delay is often interpreted as a good sign, so even when a language delay is recognised, cultural beliefs may influence the perception that this delay may influence when help is sought. Imaginative play is another example of an autistic symptom where differences in cultural norms may be related to general population samples from India, Kenya and Mexico, where less than half of the children demonstrated imaginative play. Thus, a lack of imaginative play may be a more prominent symptom of ASD in high-income Western countries than in other cultural contexts [3] (De Leeuw et al., 2020).

The prevalence of ASD has been increasing, and one explanation for this is the development and use of common diagnostic tools validated by practitioners and researchers worldwide [4]. The Centers for Disease Control and Prevention reported in 2018 that 1 in 59 children in the United States is diagnosed with ASD, with more than 90% estimated to show features of sensory disorganisation [5]. In contrast, the prevalence of ASD in European countries from 2008 to 2018 was 1/172 in Denmark, 1/125 in Norway, 1/64 in the UK, 1/806 in Portugal, 1/44, 1/175 in the Netherlands, 1/87 in Italy, 1/166 in Germany and 1/64 (3–4 years) or 1/100 (10–11 years) in Spain [4]. The way states collect and report data on ASD can vary widely. For this reason, a high prevalence of ASD can be seen in the United States due to its Autism Spectrum Disorders Surveillance System, which can collect data in greater detail and more comprehensively [5].

The fifth edition of the DSM (APA, 2013) [6,7] established new diagnostic criteria for ASD with some significant changes (see Table 1), such as the symptoms required for diagnosis being reduced from 3 to 2, the strict restrictions recommended for age of onset being removed and symptoms such as sensory hyper- or hypo-reactivity being added [6]. Criterion A of the DSM is as follows: (A) ongoing deficiencies in social communication and social interaction in different contexts, such as (A1) deficiencies in socioemotional reciprocity; (A2) deficiencies in non-verbal communicative behaviours used in social interaction; (A3) deficiencies in the development, maintenance and understanding of relationships [8].

Table 1. Evolution of the ASD diagnosis in the DSM.

Evolution of the DSM			
1980 DSM III	**1980 DSM III-R**	**1994 DSM IV**	**2013 DSM V**
Introduction of the diagnosis of autism disorder.	Subtypes such as Asperger's syndrome are added.	The term autism spectrum disorders (ASD) becomes established.	Unification of subtypes under the term autism spectrum disorder and elimination of terms such as Asperger's syndrome.

According to the deficits specified by the DSM-5 with respect to communication and social interaction, infants with ASD characteristics may manifest certain behaviours, such as failure to take turns, difficulty in maintaining adequate joint attention and complications in communication behaviour and social interaction ranging from difficulty in maintaining appropriate behaviour in play to a lack of interest in others [8,9].

Criterion B of the DSM is as follows: (B) Restrictive and repetitive patterns of behaviour, interests or activities, manifested by two or more of the following: (B1) stereotyped or repetitive movements, use of objects or speech; (B2) insistence on monotony, excessive inflexibility of routines or ritualised patterns of verbal or non-verbal behaviour; (B3) highly restricted and fixed interests that are not typical in intensity or focus of interest;

(B4) hyper- or hypo-reactivity to sensory stimuli or unusual interest in sensory aspects of the environment [10].

Children presenting deficit characteristics in restrictive and repetitive pattens of behaviour demonstrate responses that indicate restricted or repetitive interest. For example, they may not change toys or objects, and they may demonstrate atypical behaviour and impulsivity; self-injurious behaviour; repeating the same question several times; repetitive play; preoccupation with the home environment, school or daily life; an apparent indifference to pain and temperature; adverse responses to specific sounds or textures; excessive sniffing or touching of objects; or visual obsession with light or movement [10].

In addition, children with ASD are classified according to their level of functioning, i.e., grade three if they need very significant help, grade two if they need significant help and grade one if they need help; this diagnosis is established based on developmental and clinical observations of behaviour [9,11].

The severity of the clinical features of ASD may vary from subject to subject, but these features result in significant social, school and occupational underachievement [12]. Manifestations begin in infancy or early childhood, and the influence of socio-cultural factors may have an impact within the spectrum [13].

The Diagnostic and Statistical Manual of Mental Disorders (DSM-5) and its revised screening criteria for ASD aim to improve diagnostic accuracy and consistency, while emphasising the relationship of ASD with sensory processing disorder [6].

Before the DSM-5 issued its diagnostic criteria for ASD with respect to hyper- or hypo-reactivity, the aetiological implications of sensory patterns, and their relevance for assessment and intervention practices, had been reported [14]. The characteristics of sensory symptoms and the high prevalence of these symptoms in infants with ASD and how they can affect infants' exploration and interaction with the physical and social environment, as well as difficulties in regulating and organising the intensity of responses to sensory stimuli from their body or the environment, had also been described [15].

Since Ayres (1965) [16] conducted his first studies on sensory perception, different terms have been used to describe sensory integration theory [17], as well as to identify and explain concepts related to assessment and treatment. Ayres (1972) [18] defined sensory integration as the neurological process that organises sensations from one's own body and the environment, enabling effective use of the body within one's surroundings. It primarily focuses on the vestibular, proprioceptive and tactile sensory systems. Sensory integration encompasses not only dysfunction and intervention, but also their role in learning efficiency. Thus, the theory of sensory integration comprises three broad postulates. The first describes sensory integration and learning, the second defines sensory integration dysfunction and the third guides intervention [17].

Sensory processing refers to how the central nervous system and the peripheral nervous system manage information from sensory systems, which receive, modulate, integrate and organise sensory stimuli, including behavioural responses to sensory information; all of these are components of sensory processing [19]. Furthermore, sensory processing integrates information received from our sensory systems, and as part of this process, our brain interprets, selects, and organises sensory information to appropriately produce motor, behavioural, emotional and attentional responses [20].

The term sensory processing is akin to sensory integration when referring to the central nervous system's ability to process sensory information. However, the terms are not interchangeable; sensory processing is more expansive than sensory integration because sensory integration is merely a component of sensory processing [19]. In other words, sensory integration (SI) is the therapeutic treatment of sensory processing disorder, utilising play-based motor activities alongside precise challenges to influence how the child responds to sensations [21].

It is estimated that atypical sensory processing occurs in up to 90% of individuals with ASD, potentially affecting all sensory modalities [22], and that approximately 65–80% of children with ASD may experience differences in sensory reactivity [23]. Some sensory experiences for individuals with ASD can be pleasant or self-soothing, while others can be a source of irritation or discomfort [24].

Moreover, atypical sensory processing often coexists with maladaptive or problematic behaviours in children with ASD, and the presence of ASD increases the risk of experiencing social anxiety. It has been found that individuals with ASD are more prone to developing social anxiety than neurotypical individuals [25]. This relationship is even more pronounced in adults with ASD (50–70%) [26], and it also occurs in childhood and adolescence [27].

The sensory hyper-reactivity that this type of population may present has been observed to correlate with clinically elevated levels of anxiety in paediatric populations both with ASD and without ASD [23]. Research has indicated a bidirectional relationship with sensory reactivity increasing anxiety, which, in turn, increases sensory hyper-reactivity [28]. This may be because sensory hyper-reactivity contributes to avoidance, a maladaptive strategy that can further maintain anxiety [23]. Sensory hyper-reactivity in preschool children with ASD is a risk factor for the development of anxiety disorders [29]. Much of the previous research on the bidirectional relationship between anxiety and sensory hyper-reactivity in ASD has focused on sensory reactivity and the presence of anxiety. From a measurement perspective, aggregated scores on broad anxiety measures correlated with the presence of sensory hyper-reactivity. Two more recent studies investigated the relationship between atypical sensory characteristics and subtypes of anxiety [30,31]. Both studies found a relationship connecting hyper-reactivity with specific phobias and separation anxiety. Research has shown that sensory hyper-reactivity correlates with ritualism and symptoms of obsessive compulsive disorder (OCD) in adults [23]. Other research shows that increased sensory input often precedes or accompanies OCD behaviours, including repetitive behaviours and compulsions [23]. Less is known about the links between sensory reactivity differences and OCD in youth with ASD [31].

According to the nosology of sensory processing disorder proposed by Miller et al. (2007) [32], three patterns with their subtypes are included: the first pattern is sensory modulation disorder with its subtypes (hyper, hypo and sensory seeker); the second pattern corresponds to discrimination disorder; and finally, the third pattern belongs to motor-based disorder with its subtypes (dyspraxia and postural disorder). Additionally, Bundy and Lane (2020) [17] devised a schematic representing the relationship between sensory systems and manifestations of sensory integration dysfunction. From the central column outward, it depicts the processing of sensations in the CNS. Indicators of poor sensory integration and praxis are shown to the right, with behavioural consequences such as low self-efficacy and self-esteem, avoidance of participating in motor activities, poor fine motor skills and visuomotor coordination, and poor organisation and sensory seeking. To the left, the schematic shows indicators of modulation dysfunction, with atypical responses evident in sensory challenges related to attention, regulation, affect, and activity; withdrawal from and avoidance of sensory experiences; sensory seeking; and low self-efficacy and self-esteem.

The purpose of this systematic review is to examine the correlation of sensory processing disorder in children with ASD, and intervention with sensory integration as part of therapeutic treatment in children with ASD.

2. Method

A review was conducted in accordance with the guidelines and standards for systematic reviews (PRISMA) [33], in which relevant research regarding sensory integration intervention in children ASD was identified, selected and critically evaluated. A search strategy was developed using the academic search engines PubMed, Scielo, Eric, Dialnet, Springer, Base Search and Google Scholar. For this search, controlled vocabulary in both English and Spanish was employed (intervention with sensory integration in infants with ASD spectrum disorder, ASD spectrum disorder and sensory integration, sensory processing disorder in ASD), and to enhance search precision, the Boolean operators AND and OR were used. Finally, the results were restricted to studies published in English and Spanish from 2013 to 2023, yielding 16 articles meeting the inclusion criteria.

2.1. Inclusion and Exclusion Criteria

To define the eligibility criteria, the PICO strategy (Population, Intervention, Comparison, and Outcome) was employed [34], and inclusion and exclusion criteria were identified prior to proceeding with the article review. The inclusion criteria considered articles from the period ranging from 2013 to 2023. Systematic reviews and peer-reviewed articles pertaining to sensory integration intervention in children, with an emphasis on ASD, were included, provided they were published in English or Spanish. Conversely, the exclusion criteria involved articles unrelated to childhood, along with book chapters, articles under review, reports, theses, dissertations, proceedings, or works focusing solely on sensory integration, ASD, or sensory processing disorder.

2.2. Data Sources and Search Strategies

The PRISMA flowchart in Figure 1 outlines the article selection process. The classification search yielded a large number of results; initially, 1983 articles were identified, but this was reduced to 16 through a rigorous selection process that went through several stages. First, specific inclusion and exclusion criteria were used, and the time span of publication was from 2013 to 2023. In the review stage, duplicate articles and those that did not meet the criteria of the systematic review were detected and deleted. In addition, certain articles were excluded because they were not directly relevant to the research regarding the relationship of sensory integration and ASD. In the review stage, duplicate articles were detected and deleted, and certain articles that were not directly relevant to the research regarding the relationship of sensory integration and ASD were excluded, which reduced our sample to a total of 32, of which $n = 16$ were excluded due to being outside the context of intervention with sensory integration in infants with ASD. Finally, 16 articles met the inclusion criteria and were selected for systematic review.

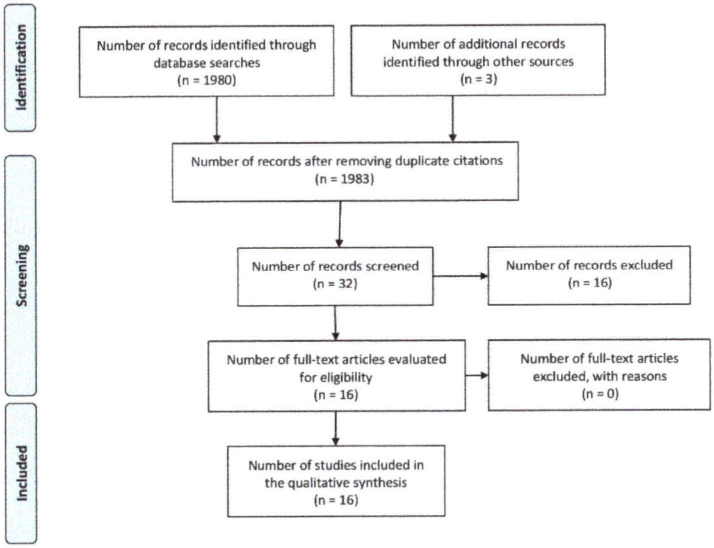

Figure 1. PRISMA diagram.

To facilitate the analysis of the 16 selected articles, a table (Table 2) was prepared including the following: (a) author identifiers, (b) year, (c) study, (d) methodological design of the study and (e) results.

Table 2. Analysis of articles.

Authors	Year	Study	Language	Method	Results
Renee Watling; Sarah Hauer [35]	2015	Effectiveness of Ayres Sensory Integration® and Sensory-Based Interventions for People With Autism Spectrum Disorder: A Systematic Review.	English	Literature published from January 2006 to April 2013 related to the effectiveness of Ayres Sensory Integration® (ASI) and sensory-based interventions (SBI) was reviewed.	Moderate evidence was found to support the use of ASI. The results of the sensory methods were mixed.
Roseann C. Schaaf; Rachel L. Dumont; Marian Arbesman; Teresa A. May-Benson [36]	2018	Efficacy of Occupational Therapy Using Ayres Sensory Integration®: A Systematic Review.	English	The review included 3 randomised controlled trials, 1 retrospective analysis and 1 single-subject ABA design, published between 2007 and 2015, that studied children with ASD.	Moderate evidence supported improvements in impairment-level outcomes of improvement in autistic behaviours and skills-based outcomes of reduced caregiver assistance with self-care activities.
Schoen SA, Lane SJ, Mailloux Z, May-Benson T, Parham LD, Smith Roley S, Schaaf RC. [37]	2019	Ayres Systematic Review of Sensory Integration for Children with Autism.	English	This study used the Council for Exceptional Children's standards to review evidence-based practice in special education in three stages.	This study evaluated the effectiveness of sensory integration research from 2006 to 2017, and determined that Ayres Sensory Integration is an evidence-based practice and is frequently requested by parents.
Citra Raditha, Setyo Handryastuti, Hardiono D Pusponegoro, Irawan Mangunatmadja [38]	2023	Positive behavioral effect of sensory integration Intervention in Young Children with Autism Spectrum disorder.	English	Non-randomised controlled trial of assessment using the Vineland Adaptive Behaviour Scale II.	A total of 72 subjects were studied and it was concluded that after the intervention communication, socialisation and daily living skills improved significantly.
Babak Kashefimehr, Hülya Kayihan, Meral Huri [39]	2018	The Effect of Sensory Integration Therapy on Occupational Performance in Children with Autism.	English	Randomised controlled trial where both the control and intervention groups were homogeneous in terms of age, and with a diagnosis of ASD grade 3.	The intervention group showed significantly greater improvement in all domains of the short child occupational profile.
Wenxin Xu, Jiwei Yao, Wenyao Liu [40]	2019	Intervention Effect of Sensory Integration Training on the Behaviors and Quality of Life of Children with Autism.	English	Randomised method. From September 2017 to December 2018, 108 participants were included in the intervention group and control group, with 54 members in each group.	Significant differences were observed after treatment, and it was concluded that the SI intervention had some effect on ASD and is of great value for the development of intervention programmes for children with ASD.
Heather M Kuhaneck, Renee Watling, Tara J Glennon [41]	2023	Ayres Sensory Integration® for Addressing Play in Autistic Children: A Multiple-Baseline Examination.	English	Non-concurrent multiple baseline design across subjects.	The findings suggest that ASI intervention can alter a child's play patterns.

Table 2. *Cont.*

Authors	Year	Study	Language	Method	Results
Elizabeth Randell, Melissa Wright, Sarah Milosevic, David Gillespie, Lucy Brookes-Howell, Monica Busse-Morris, Richard Hastings, Wakunyambo Maboshe, Rhys Williams-Thomas, Laura Mills, Renee Romeo, Nahel Yaziji, Anne Marie McKigney, Alka Ahuja, Gemma Warren, Eleni Glarou, Sue Delport, Rachel McNamara [42]	2022	Sensory Integration Therapy for Children with Autism and Sensory Processing Difficulties.	English	A parallel-group randomised controlled trial, incorporating an internal pilot and process evaluation. The trial included 138 participants, $n = 69$ per group.	The clinical effectiveness and cost-effectiveness of sensory integration therapy for children with ASD and sensory difficulties were tested.
Cristin M Holland, Erna I Blanche, Barbara L Thompson [43]	2021	Quantifying Therapists' Activities during Sensory Integration Treatment for Young Children with Autism.	English	A coding scheme that identifies specific aspects of sensory integration treatment was developed and used to analyse 34 videos of 9 children with ASD.	Quantification of therapists' actions can provide information about moment-to-moment decision making and therapist–child relationships during the daily practice of sensory integration treatment.
Ryoichiro Iwanaga, Sumihisa Honda, Hideyuki Nakane, Koji Tanaka, Haruka Toeda, Goro Tanaka [44]	2014	Pilot study: efficacy of sensory integration Therapy for Japanese Children with High-Functioning AutismSspectrum Disorder.	English	The subjects were 20 children with TEA with IQs above 70 selected from previously collected data.	Sensory integration therapy was most effective for motor coordination skills, nonverbal cognitive skills and combined sensorimotor and cognitive skills in children with ASD.
Susanne Smith Roley, Zoe Mailloux, L. Diane Parham, Roseann C. Schaaf, Christianne Joy Lane, Sharon Cermak [45]	2015	Sensory Integration and Praxis Patterns in Children With Autis.	English	Tests of sensory integration and practice (SIPT) and sensory processing measures (SPM) were conducted on clinical records of children with ASD aged 4–11 years ($n = 89$).	Children with ASD characteristically show strengths in visopraxia and difficulties with somatopraxis and vestibular functions, which appear to greatly affect participation.

Table 2. *Cont.*

Authors	Year	Study	Language	Method	Results
Włodzisław Kuliński, Adela Nowicka [46]	2020	Effects of Sensory Integration Therapy on Selected Fitness Skills in Autistic Children.	English	This study evaluated a group of 20 children (15 boys and 5 girls) aged 3–10 years. All children were diagnosed with ASD and underwent 2 years of therapy. The study assessed physical skills and their correlations with the sensory integration therapy used with the children.	Sensory integration therapy contributed to improved motor, sensory, cognitive, emotional, communicative and social development in the study patients.
Francielly Caroline Silva Costa and Luzia Iara Pfeifer [47]	2021	Sensory Integration Intervention for Children with Autistic Spectrum Disorder.	Spanish	This was a case study that was exploratory and qualitative in nature.	It was confirmed that the results of this study corroborate those found in the literature, i.e., based on sensory integration, interventions favour the improvement of various aspects of sensory processing and improved functional performance.
Hansashree Padmanabha, Pratibha Singhi, Jitendra Kumar Sahu, Prahbhjot Malhi [48]	2018	Home-based Sensory Interventions in Children with Autism Spectrum Disorder: A Randomized Controlled Trial.	English	This was a 12-week pilot, parallel-group, randomised controlled trial.	The findings suggest that sensory interventions in the home are feasible in a developing country and are suggested to have a beneficial role in ASD.
Leyla Sadat Karimi, Seyed Ebrahim Hosseini, Farzaneh Manzari-Tavakoli [49]	2017	The Effects of Sensory Integration Therapy on Physical and Motor Activity of Children with Autism Spectrum Disorders in Shiraz.	English	This descriptive study was carried out with 20 students selected in the first, second and third grades aiming to diagnose ASD in an academic setting.	Integrative therapy improves daily, organised and motor activity as well as basic and balance skills in children with ASD and may require longer treatments to improve kinesthetic skills.
Amel E. Abdel Karim, Amira H. Mohammed [50]	2015	Effectiveness of Sensory Integration Program in Motor Skills in Children with ASD.	English	Thirty-four children of both sexes with ASD (24 males and 13 females) participated in this study. They ranged in age from 40 to 65 months. The children were assessed before and after SI treatment using the Peabody Developmental Motor Scale (PDMS-2).	This study revealed a significant improvement in gross and fine motor skills. Sensory integration therapy is effective in the treatment of autistic children as it helps them to become more independent and participate in everyday activities.

3. Results

According to the analysis of the results described below (see Table 2), there are systematic reviews that link sensory integration with ASD, such as the one conducted by [35], which examined the literature related to the effectiveness of Ayres Sensory Integration (ASI) and sensory interventions [18] to improve the daily activities of infants with ASD. This study found moderate evidence suggesting that intensive, individualised ASI clinical intervention can improve functional outcomes in ASD. Three randomized controlled trials with a retrospective analysis and an ABA design concluded that there was moderate evidence of improvement in behaviour and a reduction in caregiver assistance in self-care activities [36]. Another study used the standards for evidence-based practices in special education from the Council for Exceptional Children in order to evaluate the effectiveness of sensory integration intervention for children with ASD, leading them to determine that Ayres Sensory Integration can be considered an evidence-based practice for children with ASD aged 4 to 12 years [37].

In addition to systematic reviews, there are specific research studies, detailed below in chronological order. One of the outstanding studies is a pilot study that investigated the effectiveness of sensory integration in children with high-functioning ASD. A total of 20 children were involved in the research, 8 of whom underwent personalised sensory integration treatment, and the other 12 of whom attended group therapy sessions. The findings indicate that sensory integration may be useful in improving motor, visual and visual–motor skills in preschool children with ASD, suggesting that occupational therapists and experts in the field should consider sensory integration as a valid alternative for the care of this age group [44].

In two reviewed studies, the first assessed gross and fine motor skills in 34 children with ASD (24 boys and 10 girls) before and after a sensory integration programme [50], and the second investigated the relationship between social participation and praxis and sensory integration in 89 children with ASD [45]. These studies underline the importance of sensory integration in improving motor skills and promoting social participation in children with ASD, supporting its consideration as a key component in therapeutic approaches for this group.

Another study looked at the effects of sensory integration therapy on the physical and motor activity of children with ASD. This descriptive study included 20 students from early childhood and primary education diagnosed with ASD, who received a 12-session sensory integration therapy intervention. The authors concluded that sensory integration treatment improved motor ability as well as basic and balance skills [49].

The study by Kashefimehr et al. (2018) [39] examined how sensory integration therapy impacts different areas of occupational performance by using a design that included an intervention and a control group. It was noted in the results that the intervention group showed significant improvements in all domains assessed by the short child occupational profile [39]. In another aspect, Padmanabha et al. (2019) [48] conducted a study examining a home-based sensory intervention for children with ASD, in which two groups were compared: one that received home-based sensory interventions by parents or caregivers along with standard therapy, and another group that only received standard therapy involving speech therapy and applied behaviour analysis [48]. These studies highlight the effectiveness of sensory integration, whether in clinical or home settings, and support its incorporation as a beneficial therapeutic alternative to enhance the performance and health of children with ASD.

In a study of 108 children with ASD who were divided into two groups, a sensory integration intervention group (Group A, $n = 54$) and a control group (Group B, $n = 54$), the members of the two groups were evaluated before and after treatment. Significant differences were observed after the intervention, and it was concluded that sensory integration intervention is of great value for development and intervention in programmes for children with ASD [40].

In a study to evaluate the effects of sensory integration therapy on selected physical skills in children with ASD [46], physical skills and their correlation with sensory integration therapy were evaluated in 15 boys and 5 girls. The results show that sensory integration therapy contributed to an improvement in motor, sensory, cognitive, emotional, communicative and social development in those receiving the intervention, as well as improving physical skills.

A case study involving two children and their caregivers [47] included an initial assessment followed by a six-month intervention, with a subsequent follow-up assessment. Another study [43] investigated how therapists use sensory integration therapy with children with ASD by examining the number and type of activities conducted. A coding scheme was generated in order to detect particular elements of the intervention, using recordings of nine children with ASD to analyse the performance achieved [43]. The findings indicate that sensory interactions and engagement between children and therapists are essential elements in sensory integration therapy, highlighting the importance of sensory intervention in both enhancing sensory processing and strengthening the therapeutic connection. This suggests that it has the potential to be an effective approach in the treatment of children with ASD.

Communication, socialisation and daily living skills improved significantly after sensory integration intervention in a non-randomised controlled trial of 72 infants with ASD [38], while no clinical benefit was demonstrated in a randomised controlled trial [42].

Examining a sensory integration intervention to address play in infants with ASD [41], a non-concurrent multiple baseline design was applied with three children with ASD diagnoses aged 5 and 6 years old. All three participants demonstrated changes in the frequency of specific types of play. If confirmed in future studies, sensory integration intervention could become an evidence-based treatment for improving play.

4. Discussion

This systematic review reveals several approaches in which the sensory integration reference model can be applied for the therapeutic treatment of children with ASD.

The results demonstrate that applying sensory integration can significantly improve the behaviour and daily performance of children with ASD. In addition, this structured intervention has shown improvement in communication, socialisation and daily living skills [38].

Consistency in effectiveness has been shown in the articles reviewed, as corroborated by systematic reviews concluding that Ayres sensory integration is an evidence-based practice [37] (Schoen et al., 2019). Of particular note is the study whose findings suggest that integration intervention can significantly modify children's play patterns [41].

Despite these positive findings, it is necessary to note that the heterogeneity of the interventions and the different methodologies of the studies limit our ability to generalise the effects of the interventions. Furthermore, a lack of standardisation of methods suggests that more rigorous studies are needed to confirm these results and develop clearer applications for practice in other non-clinical settings.

The results of this review support the efficacy of sensory integration in relation to previous research proposed by Watling and Hauer (2015) [35] and Schoen et al. (2019) [37]. In this regard, the results are consistent with the idea that the sensory integration approach is beneficial in improving the adaptive responses of minors with ASD.

In addition, the results of this systematic review have notable implications that impact clinical practice, and support the existing sensory integration theory in the treatment of children with ASD. They also emphasise the diversity of children with ASD, which is why the therapeutic approach needs to consider each child individually.

The results of this systematic review suggest several key contributions for occupational therapists who work with children with ASD and apply the sensory integration model in their treatment. Occupational therapists who apply this model must base their work on approaches that have demonstrated efficacy, and these interventions must be adapted to the individual needs of each child, because the response to stimuli may vary from one subject to another. Within sensory integration treatment, a wide range of adaptive responses that will benefit children with ASD should be considered, for example, including strategies that involve communication responses, social skills, psychomotor skills, etc. It is important for professionals to undertake continuous learning and undergo updated training according to the latest evidence-based methods and techniques in order to improve their skills and the effectiveness of the intervention, and guarantee the expected results.

To strengthen these findings on the efficacy of sensory integration, it is crucial to explore several areas of further research that encompass different cultural backgrounds in order to explore how treatment effects vary in different populations and to tailor interventions according to the specific needs of each group. Future studies should evaluate the persistence of sensory integration effects over the long term, including the impact of adaptive responses in school, social and family environments.

The results from this review show significant responses for the ASD spectrum disorder community who, thanks to this intervention, tend to experience significantly improvements in their quality of life, in addition to how they manage their sensory processing. This leads to greater participation in day-to-day activities that promotes better adaptation in educational and social environments. And thanks to the intervention, their families can also be included in the process. This research provides families with valuable information about which interventions are supported by scientific evidence, which will help them make informed decisions about the best treatment for their children.

This systematic review makes an important contribution, as it performs a comprehensive evaluation of interventions in a given time. This contributes to the literature and provides a clear picture of which treatments are evidence-based. In addition, it provides a critical analysis of the practices and their results, allowing professionals to select and apply treatments with a solid research base.

5. Conclusions

In conclusion, the results of this article highlight the importance of sensory integration intervention in the treatment of children with ASD. The research reviewed demonstrates that sensory integration therapy can have a positive effect in several areas in children diagnosed with ASD, including motor, cognitive, social and communication skills, as well as a significant improvement in quality of life and daily functioning in children who receive this intervention. Furthermore, sensory integration intervention for children with ASD is carried out by certified and qualified professionals, who meet the criteria for this to be considered an evidence-based practice. The results also highlight the importance of considering sensory integration intervention as an integral part of treatment programmes for children with ASD, complementing other therapeutic and educational interventions to comprehensively address these children's needs. However, the research also points to the need to address differences in outcomes, as well as the lack of success in some studies on the clinical benefits of sensory integration therapy. This highlights the importance of continuing to investigate and critically evaluate the effectiveness and applicability of this intervention in the ASD population.

In addition, the information obtained through this research shows a relationship between sensory processing and anxiety in children with ASD. However, the articles reviewed do not establish whether there was an evaluation process on the impact of anxiety prior to the sensory integration approach. Finally, most of these studies were conducted from a clinical perspective, which makes new research necessary on intervention with sensory integration from other ecological environments that are part of the daily lives of children with ASD, such as the home and school.

Author Contributions: Conceptualization, J.C.-A., M.A.R.-B., N.V.-N. and A.S.-R.; methodology, J.C.-A. and M.A.R.-B.; software, M.A.R.-B.; validation, J.C.-A., M.A.R.-B., N.V.-N. and A.S.-R.; formal analysis, J.C.-A., M.A.R.-B., N.V.-N. and A.S.-R.; investigation, J.C.-A., M.A.R.-B., N.V.-N. and A.S.-R.; resources, J.C.-A., M.A.R.-B. and N.V.-N.; data curation, J.C.-A., M.A.R.-B. and N.V.-N.; writing—original draft preparation, J.C.-A., M.A.R.-B., N.V.-N. and A.S.-R.; writing—review and editing, J.C.-A., M.A.R.-B., N.V.-N. and A.S.-R.; visualisation, J.C.-A., M.A.R.-B., N.V.-N. and A.S.-R.; supervision, M.A.R.-B. and N.V.-N.; project administration, J.C.-A., M.A.R.-B. and N.V.-N.; funding acquisition, J.C.-A. All authors have read and agreed to the published version of the manuscript.

Funding: This research received no external funding.

Institutional Review Board Statement: The study was conducted in accordance with the Declaration of Helsinki and approved by the Institutional Review Board (or Ethics Committee) of the University of Jaen (protocol code: DIC.18/19.PRY).

Informed Consent Statement: It was not necessary, the study did not involve humans.

Data Availability Statement: The data presented in this study are available on request from the corresponding author. The data are not publicly available due to ethical reasons and for confidentiality.

Conflicts of Interest: The authors declare no conflicts of interest.

References

1. Sauer, A.K.; Stanton, J.E.; Hans, S.; Grabrucker, A.M. Autism Spectrum Disorders: Etiology and Pathology. In *Autism Spectrum Disorders*; Grabrucker, A.M., Ed.; Exon Publications: Brisbane, Australia, 2021.
2. Peters, W.J.; Matson, J.L. Comparing Rates of Diagnosis Using DSM-IV-TR Versus DSM-5 Criteria for Autism Spectrum Disorder. *J. Autism Dev. Disord.* **2020**, *50*, 1898–1906. [CrossRef] [PubMed]
3. de Leeuw, A.; Happé, F.; Hoekstra, R.A. A Conceptual Framework for Understanding the Cultural and Contextual Factors on Autism Across the Globe. *Autism Res.* **2020**, *13*, 1029–1050. [CrossRef] [PubMed]
4. Provenzani, U.; Fusar-Poli, L.; Brondino, N.; Damiani, S.; Vercesi, M.; Meyer, N.; Rocchetti, M.; Politi, P. What are we targeting when we treat autism spectrum disorder? A systematic review of 406 clinical trials. *Autism* **2020**, *24*, 274–284. [CrossRef] [PubMed]
5. Kilroy, E.; Aziz-Zadeh, L.; Cermak, S. Ayres Theories of Autism and Sensory Integration Revisited: What Contemporary Neuroscience Has to Say. *Brain Sci.* **2019**, *9*, 68. [CrossRef]
6. Málaga, I.; Blanco-Lago, R.; Hedrera-Fernández, A.; Álvarez-Alvarez, N.; Oreña-Ansonera, V.A.; Baeza-Velasco, M. Prevalencia de los trastornos del espectro autista en niños en Estados Unidos, Europa y España: Coincidencias y discrepancias [Prevalence of autism spectrum disorders in USA, Europe and Spain: Coincidences and discrepancies]. *Medicina* **2019**, *79* (Suppl. 1), 4–9.
7. Yaylaci, F.; Miral, S. A Comparison of DSM-IV-TR and DSM-5 Diagnostic Classifications in the Clinical Diagnosis of Autistic Spectrum Disorder. *J. Autism Dev. Disord.* **2017**, *47*, 101–109. [CrossRef]
8. American Psychiatric Association, DSM-5 Task Force. *Diagnostic and Statistical Manual of Mental Disorders: DSM-5™*, 5th ed.; American Psychiatric Publishing, Inc.: Washington, DC, USA, 2013. [CrossRef]
9. Rando-Julve, N.; Arranz-Martínez, M.P. Didactic Proposal Aimed to Promote Communication and Social Interaction in the Student of 3 to 6 Years Old That Presents a Disorder of the Autistic Spectrum (TEA). 2020. Available online: https://zaguan.unizar.es/record/95248# (accessed on 3 June 2024).
10. Burns, C.O.; Matson, J.L. An evaluation of the clinical application of the DSM-5 for the diagnosis of autism spectrum disorder. *Expert. Rev. Neurother.* **2017**, *17*, 909–917. [CrossRef]
11. Angulo, R.F.; Zuleta, N.M.; Crissien-Quiroz, E.; Blumtritt, C. Perfil sensorial en niños con trastorno del espectro autista [Sensory profiling in children with autistic spectrum disorder]. *Arch. Venez. Farmacol. Ter.* **2020**, *39*, 105–111. [CrossRef]
12. Gutierrez, K. Identificación temprana de trastornos del espectro autista [Early identification of autism spectrum disorders]. *Acta Neurológica Colomb.* **2016**, *32*, 238–247. [CrossRef]
13. Young, H.; Oreve, M.J.; Speranza, M. Clinical characteristics and problems diagnosing autism spectrum disorder in girls. *Arch. Pediatr.* **2018**, *25*, 399–403. [CrossRef]
14. Baranek, G.T.; David, F.J.; Poe, M.D.; Stone, W.L.; Watson, L.R. Sensory Experiences Questionnaire: Discriminating sensory features in young children with autism, developmental delays, and typical development. *J. Child. Psychol. Psychiatry* **2006**, *47*, 591–601. [CrossRef] [PubMed]
15. Ben-Sasson, A.; Cermak, S.A.; Orsmond, G.I.; Tager-Flusberg, H.; Kadlec, M.B.; Carter, A.S. Sensory clusters of toddlers with autism spectrum disorders: Differences in affective symptoms. *J. Child. Psychol. Psychiatry* **2008**, *49*, 817–825. [CrossRef]
16. Ayres, J. *Patterns of Perceptual Motor Dysfuntion in Children: A Factor Analytic Study*; Southern Universities Press: Carbondale, IL, USA, 1965; pp. 365–368.
17. Bundy, A.C.; Lane, S. *Sensory Integration: Theory and Practice*, 3rd ed.; F.A. Davis: Philadelphia, PA, USA, 2020.
18. Ayres, J. *Sensory Integration and Learning Disorders*; Western Psychological Services: Torrance, CA, USA, 1972.
19. Miller, L.J.; Lane, S.J. *Sensory Integration*; American Occupational Therapy Association, Inc.: Bethesda, MD, USA, 2000; Volume 23, pp. 1–4.
20. Fernández-Pires, P.; Valera-Gran, D.; Sánchez-Pérez, A.; Hurtado-Pomares, M.; Peral-Gómez, P.; Espinosa-Sempere, C.; Juárez-Leal, I.; Navarrete-Muñoz, E.M. The Infancia y Procesamiento Sensorial (InProS-Childhood and Sensory Processing) Project:

Study Protocol for a Cross-Sectional Analysis of Parental and Children's Sociodemographic and Lifestyle Features and Children's Sensory Processing. *Int. J. Environ. Res. Public Health* **2020**, *17*, 1447. [CrossRef]
21. Randell, E.; McNamara, R.; Delport, S.; Busse, M.; Hastings, R.P.; Gillespie, D.; Williams-Thomas, R.; Brookes-Howell, L.; Romeo, R.; Boadu, J.; et al. Sensory integration therapy versus usual care for sensory processing difficulties in autism spectrum disorder in children: Study protocol for a pragmatic randomised controlled trial. *Trials* **2019**, *20*, 113. [CrossRef] [PubMed]
22. Robertson, C.E.; Baron-Cohen, S. Sensory perception in autism. *Nat. Rev. Neurosci.* **2017**, *18*, 671–684. [CrossRef] [PubMed]
23. Kirby, A.V.; Bilder, D.A.; Wiggins, L.D.; Hughes, M.M.; Davis, J.; Hall-Lande, J.A.; Lee, L.C.; McMahon, W.M.; Bakian, A.V. Sensory features in autism: Findings from a large population-based surveillance system. *Autism Res.* **2022**, *15*, 751–760. [CrossRef]
24. Uljarević, M.; Lane, A.; Kelly, A.; Leekam, S. Sensory subtypes and anxiety in older children and adolescents with autism spectrum disorder. *Autism Res.* **2016**, *9*, 1073–1078. [CrossRef]
25. Boulton, K.A.; Guastella, A.J. Social anxiety symptoms in autism spectrum disorder and social anxiety disorder: Considering the reliability of self-report instruments in adult cohorts. *Autism Res.* **2021**, *14*, 2383–2392. [CrossRef]
26. Maddox, B.B.; White, S.W. Comorbid Social Anxiety Disorder in Adults with Autism Spectrum Disorder. *J. Autism Dev. Disord.* **2015**, *45*, 3949–3960. [CrossRef]
27. Wichers, R.H.; van der Wouw, L.C.; Brouwer, M.E.; Lok, A.; Bockting, C.L.H. Psychotherapy for co-occurring symptoms of depression, anxiety and obsessive-compulsive disorder in children and adults with autism spectrum disorder: A systematic review and meta-analysis. *Psychol. Med.* **2023**, *53*, 17–33. [CrossRef]
28. Green, S.A.; Ben-Sasson, A.; Soto, T.W.; Carter, A.S. Anxiety and sensory over-responsivity in toddlers with autism spectrum disorders: Bidirectional effects across time. *J. Autism Dev. Disord.* **2012**, *42*, 1112–1119. [CrossRef] [PubMed]
29. Carpenter, K.L.H.; Baranek, G.T.; Copeland, W.E.; Compton, S.; Zucker, N.; Dawson, G. Sensory Over-Responsivity: An Early Risk Factor for Anxiety and Behavioral Challenges in Young Children. *J. Abnorm. Child. Psychol.* **2019**, *47*, 1075–1088. [CrossRef] [PubMed]
30. Black, K.R.; Stevenson, R.A.; Segers, M.; Ncube, B.L.; Sun, S.Z.; Philipp-Muller, A.; Bebko, J.M.; Barense, M.D. Linking Anxiety and Insistence on Sameness in Autistic Children: The Role of Sensory Hypersensitivity. *J. Autism Dev. Disord.* **2017**, *47*, 2459–2470. [CrossRef] [PubMed]
31. MacLennan, K.; Roach, L.; Tavassoli, T. The Relationship Between Sensory Reactivity Differences and Anxiety Subtypes in Autistic Children. *Autism Res.* **2020**, *13*, 785–795. [CrossRef]
32. Miller, L.J.; Anzalone, M.E.; Lane, S.J.; Cermak, S.A.; Osten, E.T. Concept evolution in sensory integration: A proposed nosology for diagnosis. *Am. J. Occup. Ther.* **2007**, *61*, 135–140. [CrossRef]
33. Moher, D.; Liberati, A.; Tetzlaff, J.; Altman, D.G.; PRISMA Group. Preferred reporting items for systematic reviews and meta-analyses: The PRISMA statement. *Int. J. Surg.* **2010**, *8*, 336–341. [CrossRef]
34. Frandsen, T.F.; Bruun Nielsen, M.F.; Lindhardt, C.L.; Eriksen, M.B. Using the full PICO model as a search tool for systematic reviews resulted in lower recall for some PICO elements. *J. Clin. Epidemiol.* **2020**, *127*, 69–75. [CrossRef]
35. Watling, R.; Hauer, S. Effectiveness of Ayres Sensory Integration® and Sensory-Based Interventions for People With Autism Spectrum Disorder: A Systematic Review. *Am. J. Occup. Ther.* **2015**, *69*, 6905180030p1–6905180030p12. [CrossRef]
36. Schaaf, R.C.; Dumont, R.L.; Arbesman, M.; May-Benson, T.A. Efficacy of Occupational Therapy Using Ayres Sensory Integration®: A Systematic Review. *Am. J. Occup. Ther.* **2018**, *72*, 7201190010p1–7201190010p10. [CrossRef]
37. Schoen, S.A.; Lane, S.J.; Mailloux, Z.; May-Benson, T.; Parham, L.D.; Smith Roley, S.; Schaaf, R.C. A systematic review of ayres sensory integration intervention for children with autism. *Autism Res.* **2019**, *12*, 6–19. [CrossRef]
38. Raditha, C.; Handryastuti, S.; Pusponegoro, H.D.; Mangunatmadja, I. Positive behavioral effect of sensory integration intervention in young children with autism spectrum disorder. *Pediatr. Res.* **2023**, *93*, 1667–1671. [CrossRef] [PubMed]
39. Kashefimehr, B.; Kayihan, H.; Huri, M. The Effect of Sensory Integration Therapy on Occupational Performance in Children With Autism. *OTJR Occup. Particip. Health* **2018**, *38*, 75–83. [CrossRef] [PubMed]
40. Xu, W.; Yao, J.; Liu, W. Intervention Effect of Sensory Integration Training on the Behaviors and Quality of Life of Children with Autism. *Psychiatr. Danub.* **2019**, *31*, 340–346. [CrossRef]
41. Kuhaneck, H.M.; Watling, R.; Glennon, T.J. Ayres Sensory Integration® for Addressing Play in Autistic Children: A Multiple-Baseline Examination. *Am. J. Occup. Ther.* **2023**, *77*, 7702205080. [CrossRef]
42. Randell, E.; Wright, M.; Milosevic, S.; Gillespie, D.; Brookes-Howell, L.; Busse-Morris, M.; Hastings, R.; Maboshe, W.; Williams-Thomas, R.; Mills, L.; et al. Sensory integration therapy for children with autism and sensory processing difficulties: The SenITA RCT. *Health Technol. Assess.* **2022**, *26*, 1–140. [CrossRef] [PubMed]
43. Holland, C.M.; Blanche, E.I.; Thompson, B.L. Quantifying Therapists' Activities during Sensory Integration Treatment for Young Children with Autism. *Phys. Occup. Ther. Pediatr.* **2021**, *41*, 284–299. [CrossRef]
44. Iwanaga, R.; Honda, S.; Nakane, H.; Tanaka, K.; Toeda, H.; Tanaka, G. Pilot study: Efficacy of sensory integration therapy for Japanese children with high-functioning autism spectrum disorder. *Occup. Ther. Int.* **2014**, *21*, 4–11. [CrossRef] [PubMed]
45. Roley, S.S.; Mailloux, Z.; Parham, L.D.; Schaaf, R.C.; Lane, C.J.; Cermak, S. Sensory integration and praxis patterns in children with autism. *Am. J. Occup. Ther.* **2015**, *69*, 6901220010. [CrossRef]
46. Kuliński, W.; Nowicka, A. Effects of sensory integration therapy on selected fitness skills in autistic children. *Wiad. Lek.* **2020**, *73*, 1620–1625. [CrossRef]

47. Silva Costa, F.C.; Lara Pfeifer, L. Intervención de integración sensorial en niños con trastorno del espectro autista [Sensory integration intervention for children with autistic spectrum disorder. *Revista]. Chil. Ter. Ocup.* **2021**, *22*, 197–205. [CrossRef]
48. Padmanabha, H.; Singhi, P.; Sahu, J.K.; Malhi, P. Home-based Sensory Interventions in Children with Autism Spectrum Disorder: A Randomized Controlled Trial. *Indian J. Pediatr.* **2019**, *86*, 18–25. [CrossRef] [PubMed]
49. Karimi, L.S.; Hosseini, S.E.; Manzari-Tavakoli, F. The effects ofsensory integration therapy onphysical and motor activity ofchildren with autism spectrumdisorders in shiraz. *Rep. Health Care* **2017**, *3*, 42–50. Available online: https://www.researchgate.net/publication/332963977_The_Effects_of_Sensory_Integration_Therapy_on_Physical_and_Motor_Activity_of_Children_with_Autism_Spectrum_Disorders_in_Shiraz (accessed on 31 August 2024).
50. Abdel Karim, A.E.; Mohammed, A.H. Effectiveness of sensory integration program in motor skills in children with autism. *Egypt. J. Med. Hum. Genet.* **2015**, *16*, 375–380. [CrossRef]

children

MDPI

Article

Development and Validation of the Children's Emotions Database (CED): Preschoolers' Basic and Complex Facial Expressions

Nadia Koltcheva [1,*] and Ivo D. Popivanov [1,2]

1 Department of Cognitive Science and Psychology, New Bulgarian University, 1618 Sofia, Bulgaria; ipopivanov@nbu.bg
2 Clinic of Neurology, University Hospital "Alexandrovska", 1431 Sofia, Bulgaria
* Correspondence: nkoltcheva@nbu.bg; Tel.: +359-28110414

Abstract: Background. Emotions are a crucial part of our human nature. The recognition of emotions is an essential component of our social and emotional skills. Facial expressions serve as a key element in discerning others' emotions. Different databases of images of facial emotion expressions exist worldwide; however, most of them are limited to only adult faces and include only the six basic emotions, as well as neutral faces, ignoring more complex emotional expressions. Here, we present the Children's Emotions Database (CED), a novel repository featuring both basic and complex facial expressions captured from preschool-aged children. The CED is one of the first databases to include complex emotional expressions in preschoolers. Our aim was to develop such a database that can be used further for research and applied purposes. **Methods.** Three 6-year-old children (one female) were photographed while showing different facial emotional expressions. The photos were taken under standardized conditions. The children were instructed to express each of the following basic emotions: happiness, pleasant surprise, sadness, fear, anger, disgust; a neutral face; and four complex emotions: pride, guilt, compassion, and shame; this resulted in a total of eleven expressions for each child. Two photos per child were reviewed and selected for validation. The photo validation was performed with a sample of 104 adult raters (94 females; aged 19–70 years; M = 29.9; SD = 11.40) and a limited sample of 32 children at preschool age (17 girls; aged 4–7 years; M = 6.5; SD = 0.81). The validation consisted of two tasks—free emotion labeling and emotion recognition (with predefined labels). Recognition accuracy for each expression was calculated. **Results and Conclusions.** While basic emotions and neutral expressions were recognized with high accuracy, complex emotions were less accurately identified, consistent with the existing literature on the developmental challenges in recognizing such emotions. The current work is a promising new database of preschoolers' facial expressions consisting of both basic and complex emotions. This database offers a valuable resource for advancing research in emotional development, educational interventions, and clinical applications tailored to early childhood.

Keywords: emotions; emotion recognition; Children's Emotions Database (CED); preschoolers; facial expressions; complex emotions

1. Introduction

Emotions are a crucial aspect of human nature, serving as a cornerstone of social interaction, communication, and personal development. The ability to recognize and interpret emotions accurately is an essential component of social and emotional competence. Among the various modalities through which emotions are expressed, facial expressions are especially significant for discerning others' emotional states. From infancy, humans exhibit sensitivity to facial emotional cues, highlighting the evolutionary and developmental importance of emotion recognition [1].

- **Theoretical Framework**

The importance of both basic and complex emotional expressions is supported by theoretical models such as Ekman's Neurocultural Theory of Emotion [2] and the developmental model of emotion recognition [3]. According to these frameworks, emotional expression is influenced by both biologically innate mechanisms and social learning processes. A database focused specifically on preschool-aged children's facial expressions, including complex emotions, would provide critical empirical data for testing and refining these theories. It would also shed light on the developmental trajectories of emotional competence and the interplay between innate and culturally mediated emotional processes in early childhood.

The conceptual foundation for this study rests on Ekman's Neurocultural Theory of Emotion (1972) [2], which proposes that certain basic emotions are universally expressed and recognized due to biological underpinnings, while more complex emotions require social learning. This aligns with developmental models such as Camras and Allison's (1985) [3], which suggest that while infants and toddlers can recognize and produce basic emotions, complex emotions like guilt or pride emerge only as children gain cognitive maturity and social awareness.

These complex emotions are often shaped by internalized moral codes, self-consciousness, and the understanding of social rules—processes that develop during the preschool years. Izard's Differential Emotions Theory [4] also supports this trajectory, emphasizing that emotional development is intertwined with growing cognitive and social capacities. A database reflecting these developmental shifts is thus essential for research and intervention efforts targeting early emotional competence.

- **Development of Basic Emotion Recognition in Children and Adults**

The ability to recognize facial expressions of emotion begins early in life and continues to develop throughout childhood and adolescence. Infants as young as a few months old can discriminate between happy and sad faces, and by the end of the first year, they begin to respond differentially to a wider range of emotional expressions [5,6]. During the preschool years, children's recognition accuracy for basic emotions—such as happiness, sadness, anger, and fear—improves significantly as their social experiences and language skills expand [7,8]. The recognition of emotions like disgust and surprise typically emerges later due to their more nuanced visual cues. In contrast, adults generally show high accuracy in recognizing all basic emotions, which is attributed to cumulative social learning and cognitive maturity [9,10]. However, this developmental progression is influenced by various factors, including culture, context, familiarity with the expresser, and task format [11]. Understanding these developmental trajectories is essential for creating and validating age-appropriate emotional expression databases.

Building on Darwin's foundational work on emotional expression (Darwin, 1872; 2015, as cited in [1,2]) and the identification of six universally recognized basic emotions (happiness, sadness, fear, disgust, surprise, and anger), subsequent research has consistently emphasized the centrality of facial expressions in emotional cognition and social adaptation.

- **Existing Databases**

Over the past decades, numerous facial expression databases have been developed, primarily featuring adult models and basic emotions (e.g., Ekman & Friesen, 1971, 1975 [12,13];

Tottenham et al., 2009 [14]). The availability of standardized facial expression databases has significantly advanced research on emotion recognition. These databases typically contain static images or video sequences of individuals displaying various emotions under controlled conditions. They have enabled investigations across multiple domains, ranging from basic psychological research to applications in affective computing and clinical practice [1]. However, the majority of existing databases feature adult actors and are primarily based on posed, rather than naturally elicited, expressions, thus limiting their ecological validity and relevance for developmental research.

A systematic review by Fabricio and colleagues (2022) [1] identified 36 such databases, highlighting significant heterogeneity in the methodologies used to elicit and validate emotional expressions. Most utilize adult participants whose emotional expressions are induced in semi-controlled settings to enhance spontaneity. Stimuli are predominantly static, colorful images captured against standardized backgrounds, with standardized clothing to minimize extraneous variables. Despite these efforts, substantial variation remains in the methods used to induce, capture, and validate emotional expressions, potentially affecting the comparability and generalizability of findings across studies and populations.

While facial emotion recognition has been widely studied in adults using databases like Ekman's Pictures of Facial Affect or the NimStim set [13], emotion expression databases for children—especially preschoolers—remain scarce and limited. Relying on Fabricio and colleagues' (2022) [1] systematic review, we were able to identify only five databases, some of which hardly match our target age (Table 1). The scope of all of them is restricted to basic emotions and does not include complex emotions such as pride, guilt, or compassion.

Table 1. Emotion expression databases with preschool children. (All the listed databases consist of six basic emotions (happiness, sadness, fear, disgust, anger, and surprise) and a neutral one.)

Database	Authors/Year	Age of the Models	Ethnicity	Emotions Included		Visual Characteristics/ Stimuli
				Number	Six Basic and One Neutral	
The Dartmouth Database of Children's Faces	Dalrymple et al. (2013) [15]	6–16 years	Caucasian	7	X	static, color
Picture-Set of Young Children's Affective Facial Expressions (PSYCAFE)	Franz et al. (2021) [16]	4–6 years	NA	7	X	static, color
The Child Affective Facial Expression (CAFE)	LoBue and Thrasher (2015) [17]	2–8 years	African American, Asian, Caucasian/European American, Latino, South Asian	7	X	static, color
The Child Emotion Facial Expression Set	Negrão et al. (2021) [18]	4–6 years	Caucasian, African, Asian	7	X	static, color
Child Emotions Picture Set	Romani-Sponchiado et al. (2015) [19]	6–7, 8–9, and 10–11 years	Caucasian, Afro-American, indigenous	7	X	static, gray scale

- **Need for Preschooler Databases**

A critical gap exists in the representation of children's facial expressions, particularly those of preschool-aged children. While a few databases, such as the Dartmouth Database of Children's Faces and the Picture-Set of Young Children's Affective Facial Expressions (PSYCAFE) [15,16], include young participants, they remain limited in scope, emotional range, and cultural diversity. Most notably, existing child-focused databases predominantly depict basic emotions and often overlook complex emotional states such as pride, guilt, compassion, and shame.

Understanding complex emotions is vital because they play a significant role in socialization, moral development, and interpersonal relationships in early childhood. Unlike basic emotions, complex emotions involve self-reflection, social context, and internalized standards, making them critical markers of socio-emotional and cognitive maturity. Their accurate recognition and expression are associated with school readiness, peer acceptance, and long-term mental health outcomes. The lack of available stimuli depicting complex emotions among preschool-aged children thus impedes a comprehensive understanding of early emotional development and limits the effectiveness of diagnostic and educational interventions.

The development of the Children's Emotions Database (CED) was motivated by the limitations of existing emotional expression sets. While widely used databases such as the Pictures of Facial Affect (POFA; [13]) and NimStim [14] offer well-controlled images of adult models displaying basic emotions, they are not suitable for studies involving children—either as expressers or as target perceivers. Even among child-focused databases, including the Dartmouth Database [15], the Child Affective Facial Expression set (CAFE; [17]), and Child Emotions Picture Set [19], consist of expressions of basic emotions such as happiness, sadness, anger, fear, and surprise, and predominantly include school-aged children or adolescents.

This presents a significant developmental gap, as few databases focus specifically on preschool-aged children, despite this being a key period for the acquisition of emotional understanding, vocabulary, and regulation strategies. Children between the ages of 3 and 7 experience rapid growth in their socio-emotional competence, yet their emotional expressions—especially more complex, socially constructed emotions—are underrepresented in available resources. Most existing databases lack depictions of nuanced emotional states such as guilt, pride, or compassion, which are essential for studying the emergence of moral reasoning, empathy, and interpersonal behavior in early childhood.

The CED was specifically designed to address these limitations by including a diverse sample of preschoolers and capturing both basic and complex emotional expressions. By introducing a validated set of facial expressions across a broader emotional range, the CED provides a novel tool for researchers, educators, and clinicians to investigate emotion recognition, regulation, and expression in early childhood with greater sensitivity and developmental relevance.

- **Aim of the Present Work**

In response to these gaps, the current project aimed to develop and validate the Children's Emotions Database (CED), a novel repository of facial expressions featuring both basic and complex emotions in preschool-aged children. By addressing key limitations of existing databases, the CED offers a valuable resource for researchers, educators, and clinicians. It holds the potential to enhance the ecological validity of studies in emotional development, support the creation of developmentally appropriate assessment tools, and inform interventions aimed at fostering children's social and emotional skills.

2. Materials and Methods

2.1. Participants in the Database

Three 6–7-year-old children (M = 6.86; SD = 0.29) of Bulgarian origin (two boys and one girl) took part in the photoshoot. All children were members of theater school for at least one year. They participated voluntarily and their parents signed an informed consent form. The study was approved by the Ethical Committee of the Department of Cognitive Science and Psychology, NBU (No. 647/12 September 2023).

2.2. Stimuli Development Procedure

Frontal color portrait photos of the children were taken with a Sony ZV-1 20MPx digital camera under standardized conditions (see Figure 1). The camera was positioned approximately 1.5 m away from the actor. Two 1000 W UNOMAT LX901GZ video-lamps were placed at each side of the actors, ensuring uniform light distribution on the face. All children were photographed against the same white vinyl background, placed 50 cm behind the actor.

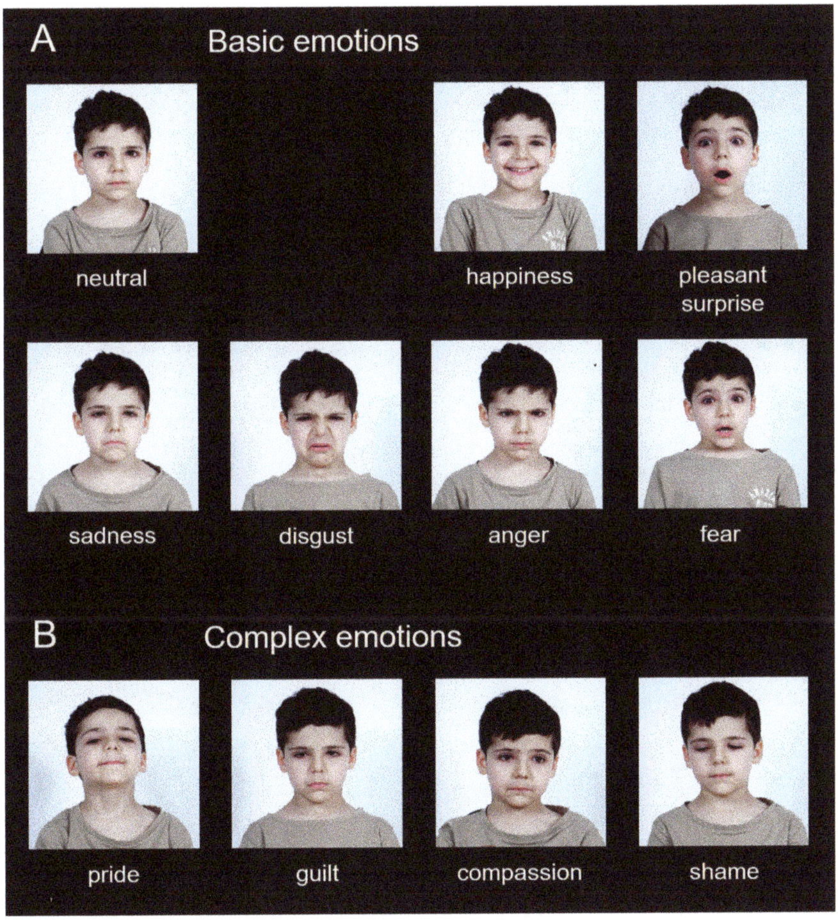

Figure 1. Examples of the facial emotional expressions of one of the children in CED. (**A**) Six basic facial emotional expressions together with a neutral facial expression. (**B**) Four complex facial emotional expressions.

After a brief general instruction session, the children were asked to express each of the following 11 emotional states with their faces: joy, surprise (pleasant), sadness, fear, anger, disgust, a neutral face (basic emotions, Figure 1A), and pride, guilt, compassion, and shame (complex emotions, Figure 1B). When necessary, short situational stories were used to elicit more authentic expressions, e.g., *Imagine you have seen your best friend suffer because his pet is sick* (compassion) or *Imagine you have received a fantastic present for your birthday* (pleasant surprise).

Several photos were taken for each emotional state for each child. Then only a few photos (1–3) per child per emotion were selected by two experts resulting in total of 60 photos (6 stimuli of happiness, 5 of pleasant surprise, 6 of pride, 5 of neutral expression, 6 of compassion, 6 of anger, 6 of fear, 6 of sadness, 5 of disgust, 6 of shame, and 3 of guilt). After the photos were taken, they were cropped and saved in JPEG format with a size of 1050 × 1050 px, a resolution of 96 dpi, and a bit depth of 24. As such, the images were ready for online validation procedure.

2.3. Validation of the Children's Emotions Database with Adult Subjects

2.3.1. Participants for Validation

In total, 104 Bulgarian volunteers (94 women), aged between 19 and 70 years (M = 29.9 SD = 11.40), took part in the online validation. Some of them were students at NBU, who participated in exchange for a course credit, and others were volunteers recruited through social networks.

2.3.2. Validation Procedure

The validation procedure took place online using the Google Forms platform. The participants had to perform two separate tasks with all the photos. In both tasks, the photos were shown one at a time in a randomized order.

The first task was the free naming of the emotion depicted in the picture. The participants were allowed to give a short answer (a word or a phrase) in a free written form. The second task was an emotion recognition task with 11 predefined labels—the same as the ones shown in Figure 1.

- **Task 1: Free Emotion Naming**
  Participants were instructed to provide a free-text label (a word or short phrase) that best described the emotion depicted in each photograph.
- **Task 2: Emotion Recognition with Predefined Labels**
  Participants were asked to identify the emotion displayed in each photograph by selecting from 11 predefined emotion labels corresponding to the intended expressions (i.e., the 7 basic and 4 complex emotions).

Each task was completed independently to avoid bias. No feedback was provided between tasks.

2.4. Validation of the Children's Emotions Database with Preschool Children Subjects

2.4.1. Participants for Validation

We were able to collect responses for the free naming and emotion recognition task from a limited sample of preschool children of Bulgarian origin. In total, 32 of them (17 girls), aged between 4 years and 9 months and 7 years and 11 months (M = 6 years and 6 months; SD = 9.7 months), took part in the online free naming task. In total, 28 of them (15 girls), aged between 4 years and 11 months and 7 years and 11 months (M = 6 years and 10 months SD = 8.6 months), also took part in the online recognition task. The children's parents filled in an informed consent form for their participation in the study.

2.4.2. Validation Procedure

The validation procedure took place online using the Google Forms platform. The participants had to perform the same two tasks—free emotion naming and emotion recognition with predefined labels—as described in Section 2.3.2. Since most of the participants were not able to read, their parents read the instructions and the possible choices (in the recognition task) and filled in their answers.

Each task was completed independently, with a short break between them. No feedback was provided between tasks.

3. Results

To evaluate the accuracy of emotion naming and recognition across age groups, we used descriptive statistics (mean percent correct responses) and inferential analyses. For comparisons between adult and child participants, we performed a series of 2 × 2 Chi-square (χ^2) tests on each emotional expression (group: adults vs. children × response: correct vs. incorrect). In addition, we compared confusion patterns using confusion matrices.

3.1. Accuracy of Emotion Naming and Recognition—Adults

Recognition accuracy for each emotional expression was assessed separately for the free naming and predefined-label tasks.

For the free naming task, two experts (the co-authors) reviewed all the answers. The answers were then classified as *correct* (correctly named emotion or its synonym) or *incorrect* (incorrectly named emotion). The percent correct responses for each stimulus were counted and then averaged across the 11 emotional expressions (see column "Naming" in Table A1 and white bars in Figure 2).

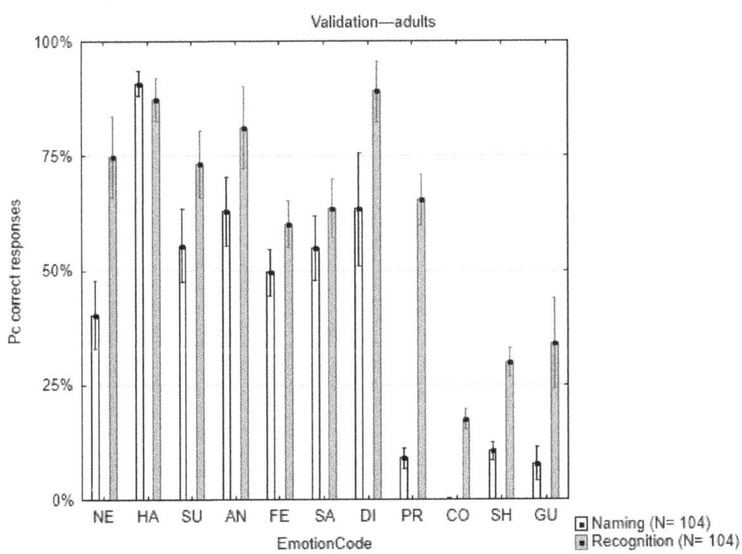

Figure 2. Averaged percent correct responses for each of the eleven facial emotional expressions in the free naming (white bars) and the recognition (gray bars) validation tasks with adult participants. Note the relatively lower correct responses for the complex emotions. Vertical bars denote SEM. Emotion codes: HA—Happiness; SU—Surprise; PR—Pride; NE—Neutral; CO—Compassion; AN—Anger; FE—Fear; SA—Sadness; DI—Disgust; SH—Shame; GU—Guilt.

For the emotion recognition task, first the correct answers were counted for each stimulus. Next, the percent correct responses were averaged across the 11 emotional expressions. The results are summarized in Table A1 (column "Recognition") and visualized in Figure 2 (gray bars).

All basic emotional expressions (happiness, surprise, sadness, fear, anger, and disgust) were recognized and named correctly by more than 50% of the participants. The only exception was the naming of the neutral expression, which was a bit lower than 50% correct. As expected, the percent correct responses were relatively higher in the recognition task for all emotional expressions (except for happiness). This was especially evident for the complex emotional expressions (pride, guilt, compassion, and shame) which had very low percent correct responses in the naming task.

3.2. Accuracy of Emotion Naming and Recognition—Children

The percent correct responses were calculated following an identical procedure to the one described in Section 3.1. Figure 3 (and Table A2) presents the results of both tasks.

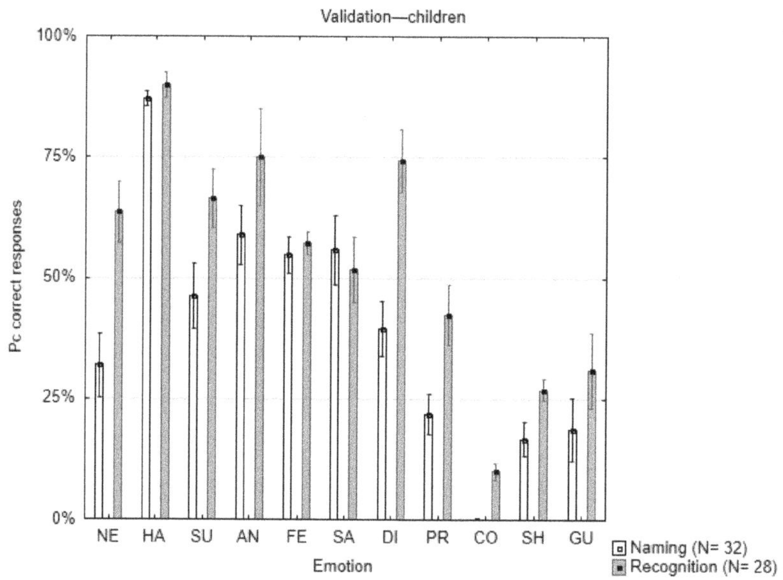

Figure 3. Averaged percent correct responses for each of the eleven facial emotional expressions in the free naming (white bars) and the recognition (gray bars) validation tasks with child participants. Note the relatively lower correct responses for the complex emotions. Vertical bars denote SEM. Emotion codes: same as in Figure 2.

3.3. Comparison Between Naming and Recognition Performance of Adults and Children

Adults' and children's accuracy in naming and recognizing emotions was compared by means of multiple 2 × 2 (adults vs. children × correct vs. incorrect responses) Chi-square tests for each of the 11 facial emotional expressions. Bonferroni correction for multiple comparisons was applied to the significance level, so that alpha was set to 0.005 (0.05/11).

The analysis revealed no statistically significant differences between adults and children in either the naming or recognition tasks for any of the individual emotional expressions (all p-values > 0.02). These results suggest that preschool-aged children are able to name and recognize both basic and some complex emotional expressions with a similar level of accuracy as adults. However, it is important to note that the sample size for the children's group was limited, which may affect the power of the statistical tests and should be considered when interpreting the results.

3.4. Confusion Patterns Among Emotions—Adults and Children

Furthermore, we evaluated which of the emotional expressions were confused most often. To do so, we counted the percent correct responses for each expression as well as the percent responses to other expressions. Thus, we created confusion matrices with the presented expressions and responses given by the adult and child participants in the recognition task (Figure 4A,B and Tables A3 and A4, respectively). These matrices present the proportion of responses for each presented emotion and reveal common patterns of confusion.

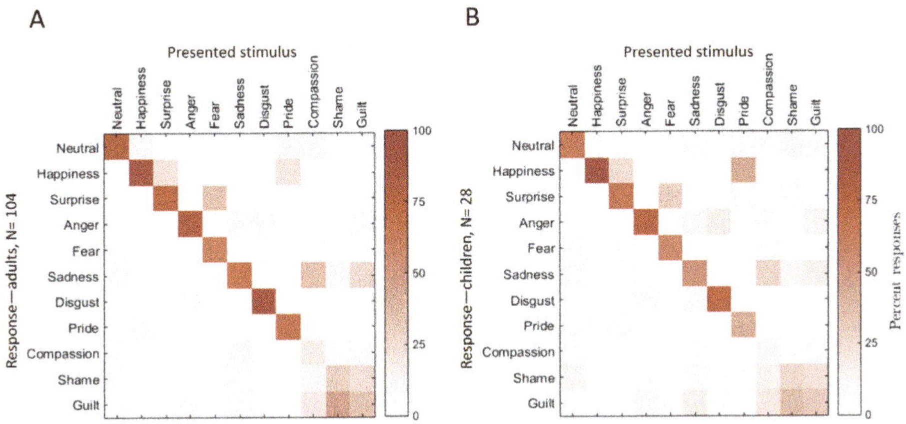

Figure 4. Confusion matrices based on the responses in the emotion recognition task of adult (**A**) and child participants (**B**). Each column represents 1 of the 11 presented emotional categories. Each row represents the responses out of the 11 possible labels. Note that complex emotions were confused much more than the basic ones.

In Figure 4, it can be noted that in both tested samples, basic emotions and neutral expressions were recognized more reliably than the complex ones, as the latter ones were often confused with other expressions. For example, the expression of *compassion* was recognized more often as "sadness" and almost equally often as "guilt" in comparison to the correct response "compassion". Similarly, expressions of *shame* and *guilt* were often confused between each other and also with the basic expression of "sadness".

Additionally, our data replicates the well-known finding that the expression of *fear* can often be confused with "surprise". Interestingly, the emotional expression of *surprise* was also recognized sometimes as "happiness".

Finally, the confusion matrices of adults and children (Figure 4A,B, respectively) look remarkably similar, showing that children at preschool age are able to recognize basic and some complex emotions as well as adults. Thus, the CED seems to be a valid instrument for testing emotion recognition in preschool age.

Overall, basic emotions were recognized more reliably, showing higher percentages along the matrix diagonal (correct responses) compared to complex emotions. Complex emotions were frequently confused with both other complex emotions and some basic emotions.

4. Discussion and Conclusions

The present study introduces the Children's Emotions Database (CED), a novel and valuable resource that addresses a significant gap in the field of developmental psychology. The CED captures both basic and complex emotional facial expressions in preschool-aged children, offering a more ecologically valid and developmentally appropriate tool than existing facial expression databases. By providing a validated set of emotional expressions from young children, the CED offers new opportunities for research, educational programming, and clinical assessments related to early childhood emotional development.

Our findings demonstrate that adults are more reliably able to recognize basic emotions, such as happiness, sadness, and anger, compared to complex emotions like pride, guilt, compassion, and shame. This pattern is consistent with established research in emotion recognition and emotional development, highlighting the developmental and cognitive challenges associated with understanding nuanced emotional states. The difficulty in recognizing complex emotions reflects the socially embedded nature of these feelings, emphasizing the need for further research to explore the developmental trajectory of emotion recognition in young children. Confusion matrices further indicated that participants tended to confuse complex emotions with each other or with related basic emotions such as sadness. These results underscore the inherent challenges associated with recognizing nuanced emotional states, especially when expressed by young children.

4.1. Theoretical Implications

These findings align with developmental theories suggesting that the ability to express and recognize basic emotions emerges early and is largely biologically driven [5], whereas the understanding of complex emotions develops later and is influenced more heavily by socialization and cultural learning [6]. The difficulty observed in recognizing complex emotions in preschool-aged children supports the notion that such emotions are still emerging during early childhood, both in their expression and their recognition by others.

Furthermore, the systematic confusion between certain emotions, such as fear and surprise, corroborates long-standing observations in emotion research, emphasizing that perceptual overlaps between certain expressions pose recognition challenges even for adults.

The relatively low recognition rates for complex emotions, such as compassion, likely reflect their higher cognitive demands. These emotions are socially constructed and involve internal states that may not manifest through prototypical facial expressions. This aligns with theories that suggest these emotions require mentalizing skills, such as perspective-taking.

Moreover, recognition accuracy was higher when participants selected from predefined categories compared to the free labeling task, suggesting that top–down processes (e.g., category priming) play a role in disambiguating children's expressions. This has implications for designing emotion recognition tasks and underscores the importance of training both children and adults in recognizing complex emotional cues.

4.2. Practical Applications

The CED offers significant potential applications for research, education, and clinical practice. In research contexts, the database can facilitate more ecologically valid investigations into the development of emotion recognition abilities in both typical and atypical populations (e.g., children with autism spectrum disorder). In educational settings, the CED can inform the design of social–emotional learning (SEL) programs that aim to improve children's emotional literacy by using age-relevant and realistic stimuli. Clinically, the database may aid in early identification of socio-emotional difficulties and inform intervention programs targeting emotional understanding and regulation.

4.3. Limitations

Several limitations of the present study warrant acknowledgment. The validation sample was predominantly female (approximately 91%), which may have influenced recognition patterns. Although emotion recognition is generally robust across genders, some research suggests subtle differences that future studies should examine more closely. The emotional expressions were elicited through instructed posing rather than spontaneous emotional reactions, which, although standardized, might not fully capture the dynamic and nuanced nature of natural emotional expression.

4.4. Future Directions

Future research should aim to expand the CED by testing emotion recognition among children, not only adults, which would provide important insights into developmental trajectories of emotional understanding. Investigating the impact of context (e.g., body language and situational background) alongside facial expressions would further enrich the ecological validity of emotion recognition research in young populations.

Future validation efforts should include child raters to assess developmental changes in peer emotion recognition. Peer-based evaluations are critical for educational and social intervention research and would improve the ecological validity of the database.

Additionally, cross-cultural validation is an essential next step. Collaborative studies across different cultural groups would allow the universality of both basic and complex emotion recognition to be tested. These expansions would strengthen the database's utility across diverse populations and contexts.

4.5. Conclusions

The Children's Emotions Database (CED) addresses a significant gap in the field of developmental psychology by providing a validated set of facial emotional expressions from preschool-aged children, including both basic and complex emotions. The CED represents an important advancement in the study of early emotional development. It provides a unique and promising tool for researchers, educators, and clinicians interested in understanding and supporting young children's social and emotional growth. By offering a more comprehensive set of emotional expressions, the CED has the potential to inform the development of age-appropriate emotion recognition assessments, contribute to more effective social and emotional learning (SEL) programs, and aid in the early detection of developmental anomalies. As the database expands and future work addresses its limitations, it will continue to enrich our understanding of emotional development in early childhood, paving the way for advancements in emotion research and interventions.

Author Contributions: Conceptualization, N.K. and I.D.P.; methodology, N.K. and I.D.P.; formal analysis, I.D.P.; investigation, N.K. and I.D.P.; data curation, N.K. and I.D.P.; writing—original draft preparation, N.K. and I.D.P.; writing—review and editing, N.K. and I.D.P.; visualization, I.D.P. All authors have read and agreed to the published version of the manuscript.

Funding: This research received no external funding.

Institutional Review Board Statement: This study was conducted in accordance with the Declaration of Helsinki and approved by the Ethics Committee of the "Cognitive Science and Psychology" Department, New Bulgarian University (No. 647/12 September 2023).

Informed Consent Statement: Parental informed consent was obtained for all subjects involved in this study.

Data Availability Statement: The data presented in this study are available on request from the corresponding author. The data are not publicly available due to the privacy.

Acknowledgments: The authors express their gratitude to Liudmila Dobreva, Albena Liubomirova, and Teodora Georgieva, students from the Master's Program "Developmental Psychology", "Cognitive Science and Psychology" Department, New Bulgarian University, for their efforts, assistance, and contribution.

Conflicts of Interest: The authors declare no conflicts of interest.

Abbreviations

The following abbreviations are used in this manuscript:

CED	Children's Emotions Database
HA	Happiness
SU	Surprise
PR	Pride
NE	Neutral
CO	Compassion
AN	Anger
FE	Fear
SA	Sadness
DI	Disgust
SH	Shame
GU	Guilt
SEM	Standard Error of Mean

Appendix A

The accuracy data from the two validation tasks with adult subjects are presented in Table A1.

Table A1. Proportion of correct responses in free naming task and emotion recognition task with predefined labels for each stimulus and averaged across emotion category. The data is based on the answers of N = 104 adult participants.

		Naming			Recognition		
Stimulus	Code	Pc Correct	M(Pc Correct)	SE(Pc Correct)	Pc Correct	M(Pc Correct)	SE(Pc Correct)
1_M1_HA_040	HA	0.90	0.91	0.03	0.74	0.87	0.05
1_M1_HA_056	HA	0.95			0.97		
1_M2_HA_282	HA	0.98			0.81		
1_M2_HA_281	HA	0.95			0.76		
1_F1_HA_161	HA	0.82			0.99		
1_F1_HA_162	HA	0.84			0.96		
1_MA_SU_094	SU	0.48	0.63	0.08	0.67	0.81	0.09
1_F1_SU_200	SU	0.65			0.91		
1_F1_SU_201	SU	0.41			0.64		
1_M2_SU_338	SU	0.40			0.89		
1_M2_SU_313	SU	0.82			0.54		

Table A1. *Cont.*

		Naming			Recognition		
Stimulus	**Code**	**Pc Correct**	**M(Pc Correct)**	**SE(Pc Correct)**	**Pc Correct**	**M(Pc Correct)**	**SE(Pc Correct)**
1_M1_PR_128	PR	0.04	0.09	0.02	0.45	0.65	0.06
1_M1_PR_130	PR	0.17			0.84		
1_F1_PR_208	PR	0.07			0.56		
1_F1_PR_209	PR	0.06			0.63		
1_M2_PR_357	PR	0.07			0.71		
1_M2_PR_356	PR	0.13			0.74		
1_M1_NE_045	NE	0.56	0.55	0.07	0.92	0.63	0.06
1_M1_NE_049	NE	0.45			0.81		
1_F1_NE_159	NE	0.45			0.81		
1_M2_NE_379	NE	0.43			0.80		
1_M2_NE_280	NE	0.13			0.40		
1_M1_CO_151	CO	0.00	0.63	0.12	0.12	0.89	0.07
1_M1_CO_148	CO	0.00			0.14		
1_F1_CO_270	CO	0.00			0.22		
1_F1_CO_273	CO	0.00			0.26		
1_M2_CO_395	CO	0.00			0.16		
1_M2_CO_396	CO	0.00			0.14		
1_M1_AN_063	AN	0.63	0.55	0.08	0.78	0.73	0.07
1_M1_AN_058	AN	0.35			0.39		
1_F1_AN_167	AN	0.60			0.82		
1_F1_AN_177	AN	0.55			0.92		
1_M2_AN_291	AN	0.88			0.99		
1_M2_AN_284	AN	0.78			0.96		
1_M1_FE_070	FE	0.38	0.11	0.02	0.54	0.30	0.03
1_M1_FE_071	FE	0.36			0.42		
1_F1_FE_181	FE	0.62			0.74		
1_F1_FE_182	FE	0.64			0.69		
1_M2_FE_305	FE	0.42			0.53		
1_M2_FE_304	FE	0.56			0.68		
1_M1_SA_090	SA	0.74	0.50	0.05	0.75	0.60	0.05
1_F1_SA_193	SA	0.39			0.61		
1_F1_SA_198	SA	0.72			0.82		
1_F1_SA_251	SA	0.38			0.41		
1_M2_SA_312	SA	0.40			0.50		
1_M2_SA_310	SA	0.64			0.72		

Table A1. *Cont.*

Stimulus	Code	Naming			Recognition		
		Pc Correct	M(Pc Correct)	SE(Pc Correct)	Pc Correct	M(Pc Correct)	SE(Pc Correct)
1_M1_DI_109	DI	0.68	0.00	0.00	0.63	0.17	0.02
1_M1_DI_100	DI	0.20			0.94		
1_F1_DI_170	DI	0.58			0.90		
1_M2_DI_328	DI	0.77			0.98		
1_M2_DI_334	DI	0.93			0.99		
1_M1_SH_146	SH	0.10	0.08	0.04	0.32	0.34	0.10
1_M1_SH_134	SH	0.13			0.31		
1_F1_SH_211	SH	0.02			0.18		
1_F1_SH_221	SH	0.11			0.29		
1_M2_SH_369	SH	0.13			0.42		
1_M2_SH_367	SH	0.15			0.28		
1_M1_GU_144	GU	0.09	0.40	0.07	0.38	0.75	0.09
1_F1_GU_220	GU	0.13			0.49		
1_M2_GU_364	GU	0.01			0.15		

Appendix B

The accuracy data from the two validation tasks with children subjects are presented in Table A2.

Table A2. Proportion of correct responses in free naming task and emotion recognition task with predefined labels for each stimulus and averaged across emotion category. The data for the naming task is based on the answers of N = 32 children and the data for the recognition task is based on the answers of N = 28 children.

Stimulus	Code	Naming			Recognition		
		Pc Correct	M(Pc Correct)	SE(Pc Correct)	Pc Correct	M(Pc Correct)	SE(Pc Correct)
1_M1_HA_040	HA	0.81	0.87	0.01	0.79	0.90	0.03
1_M1_HA_056	HA	0.91			0.96		
1_M2_HA_282	HA	0.84			0.89		
1_M2_HA_281	HA	0.88			0.93		
1_F1_HA_161	HA	0.91			0.93		
1_F1_HA_162	HA	0.88			0.89		
1_MA_SU_094	SU	0.41	0.59	0.06	0.57	0.75	0.10
1_F1_SU_200	SU	0.63			0.79		
1_F1_SU_201	SU	0.44			0.64		
1_M2_SU_338	SU	0.25			0.82		
1_M2_SU_313	SU	0.59			0.50		

Table A2. *Cont.*

		Naming			Recognition		
Stimulus	**Code**	**Pc Correct**	**M(Pc Correct)**	**SE(Pc Correct)**	**Pc Correct**	**M(Pc Correct)**	**SE(Pc Correct)**
1_M1_PR_128	PR	0.09	0.22	0.04	0.21	0.42	0.06
1_M1_PR_130	PR	0.34			0.61		
1_F1_PR_208	PR	0.19			0.39		
1_F1_PR_209	PR	0.13			0.29		
1_M2_PR_357	PR	0.31			0.50		
1_M2_PR_356	PR	0.25			0.54		
1_M1_NE_045	NE	0.34	0.56	0.07	0.71	0.52	0.07
1_M1_NE_049	NE	0.44			0.75		
1_F1_NE_159	NE	0.38			0.64		
1_M2_NE_379	NE	0.38			0.68		
1_M2_NE_280	NE	0.06			0.39		
1_M1_CO_151	CO	0.00	0.39	0.06	0.04	0.74	0.06
1_M1_CO_148	CO	0.00			0.14		
1_F1_CO_270	CO	0.00			0.11		
1_F1_CO_273	CO	0.00			0.11		
1_M2_CO_395	CO	0.00			0.14		
1_M2_CO_396	CO	0.00			0.07		
1_M1_AN_063	AN	0.66	0.46	0.07	0.68	0.66	0.06
1_M1_AN_058	AN	0.31			0.29		
1_F1_AN_167	AN	0.56			0.86		
1_F1_AN_177	AN	0.63			0.86		
1_M2_AN_291	AN	0.75			0.93		
1_M2_AN_284	AN	0.63			0.89		
1_M1_FE_070	FE	0.50	0.17	0.03	0.50	0.27	0.02
1_M1_FE_071	FE	0.53			0.54		
1_F1_FE_181	FE	0.72			0.64		
1_F1_FE_182	FE	0.56			0.57		
1_M2_FE_305	FE	0.53			0.54		
1_M2_FE_304	FE	0.44			0.64		
1_M1_SA_090	SA	0.75	0.55	0.04	0.68	0.57	0.02
1_F1_SA_193	SA	0.50			0.50		
1_F1_SA_198	SA	0.72			0.64		
1_F1_SA_251	SA	0.47			0.21		
1_M2_SA_312	SA	0.28			0.50		
1_M2_SA_310	SA	0.63			0.57		

Table A2. *Cont.*

Stimulus	Code	Naming			Recognition		
		Pc Correct	M(Pc Correct)	SE(Pc Correct)	Pc Correct	M(Pc Correct)	SE(Pc Correct)
1_M1_DI_109	DI	0.44	0.00	0.00	0.57	0.10	0.02
1_M1_DI_100	DI	0.25			0.68		
1_F1_DI_170	DI	0.28			0.68		
1_M2_DI_328	DI	0.44			0.89		
1_M2_DI_334	DI	0.56			0.89		
1_M1_SH_146	SH	0.13	0.19	0.07	0.29	0.31	0.08
1_M1_SH_134	SH	0.16			0.36		
1_F1_SH_211	SH	0.03			0.21		
1_F1_SH_221	SH	0.19			0.29		
1_M2_SH_369	SH	0.28			0.21		
1_M2_SH_367	SH	0.22			0.25		
1_M1_GU_144	GU	0.28	0.32	0.07	0.21	0.64	0.06
1_F1_GU_220	GU	0.22			0.46		
1_M2_GU_364	GU	0.06			0.25		

Appendix C

The confusion matrices of the emotion stimuli and all possible responses in the emotion recognition task with adult and child participants are shown in Tables A3 and A4, respectively.

Table A3. Confusion matrix of the responses in the emotion recognition task of N = 104 adult participants [1].

Adults		Presented Stimuli										
		HA	SU	PR	NE	CO	AN	FE	SA	DI	SH	GU
Responses	HA	87%	22%	22%	0%	0%	0%	0%	0%	1%	0%	0%
	SU	0%	73%	0%	0%	1%	1%	35%	0%	0%	3%	0%
	PR	5%	2%	67%	2%	1%	1%	0%	2%	0%	0%	1%
	NE	8%	0%	10%	75%	8%	0%	0%	1%	0%	2%	2%
	CO	0%	0%	0%	3%	18%	1%	1%	9%	0%	5%	5%
	AN	0%	0%	0%	2%	1%	81%	0%	8%	6%	0%	5%
	FE	0%	2%	0%	4%	3%	1%	60%	1%	0%	2%	1%
	SA	0%	0%	0%	5%	37%	3%	0%	65%	1%	11%	27%
	DI	0%	0%	0%	0%	1%	5%	1%	2%	89%	1%	1%
	SH	0%	0%	0%	3%	10%	2%	0%	4%	1%	30%	24%
	GU	0%	0%	0%	3%	20%	5%	2%	7%	1%	47%	35%

[1] The basic emotions are highlighted in gray. Each column represents 1 of the 11 presented emotional categories. Each column represents the responses out of the 11 possible labels. The basic emotional expressions are highlighted in gray.

Table A4. Confusion matrix of the responses in the emotion recognition task of N = 28 child participants [1].

Children		Presented Stimuli										
		HA	SU	PR	NE	CO	AN	FE	SA	DI	SH	GU
Responses	HA	90%	25%	44%	1%	1%	0%	2%	1%	1%	1%	0%
	SU	2%	66%	4%	1%	1%	1%	31%	0%	0%	5%	0%
	PR	4%	4%	42%	4%	2%	0%	0%	2%	0%	0%	0%
	NE	3%	1%	7%	64%	7%	1%	0%	4%	0%	10%	5%
	CO	0%	0%	1%	4%	10%	1%	1%	3%	0%	3%	7%
	AN	0%	0%	0%	2%	7%	75%	0%	10%	18%	2%	14%
	FE	0%	4%	1%	3%	7%	1%	57%	1%	0%	1%	1%
	SA	0%	0%	0%	6%	29%	2%	2%	52%	0%	14%	17%
	DI	0%	0%	1%	1%	0%	5%	2%	6%	74%	2%	2%
	SH	1%	0%	1%	13%	15%	4%	1%	5%	3%	27%	23%
	GU	0%	0%	0%	3%	21%	10%	4%	17%	4%	36%	31%

[1] The basic emotions are highlighted in gray. Each column represents 1 of the 11 presented emotional categories. Each column represents the responses out of the 11 possible labels. The basic emotional expressions are highlighted in gray.

References

1. Fabrício, D.d.M.; Ferreira, B.L.C.; Maximiano-Barreto, M.A.; Muniz, M.; Chagas, M.H.N. Construction of face databases for tasks to recognize facial expressions of basic emotions: A systematic review. *Dement. Neuropsychol.* **2022**, *16*, 388–410. [CrossRef] [PubMed]
2. Ekman, P. *Universals and Cultural Differences in Facial Expressions of Emotion. Nebraska Symposium on Motivation*; Cole, J., Ed.; University of Nebraska Press: Lincoln, NE, USA, 1972; Volume 19, pp. 207–283.
3. Camras, L.A.; Allison, K. Children's understanding of emotional facial expressions and verbal labels. *J. Nonverbal Behav.* **1985**, *9*, 84–94. [CrossRef]
4. Izard, C.E. Basic emotions, relations among emotions, and emotion-cognition relations. *Psychol. Rev.* **1992**, *99*, 561–565. [CrossRef] [PubMed]
5. Serrano, J.M.; Iglesias, J.; Loeches, A. Visual discrimination and recognition of facial expressions of anger, fear, and surprise in 4- to 6-month-old infants. *Dev. Psychobiol.* **1992**, *25*, 411–425. [CrossRef] [PubMed]
6. Walker-Andrews, A.S. Infants' perception of expressive behaviors: Differentiation of multimodal information. *Psychol. Bull.* **1997**, *121*, 437–456. [CrossRef] [PubMed]
7. Widen, S.C.; Russell, J.A. A closer look at preschoolers' freely produced labels for facial expressions. *Dev. Psychol.* **2003**, *39*, 114–128. [CrossRef] [PubMed]
8. Herba, C.; Phillips, M. Annotation: Development of facial expression recognition from childhood to adolescence: Behavioural and neurological perspectives. *J. Child Psychol. Psychiatry* **2004**, *45*, 1185–1198. [CrossRef] [PubMed]
9. Ruffman, T.; Henry, J.D.; Livingstone, V.; Phillips, L.H. A meta-analytic review of emotion recognition and aging: Implications for neuropsychological models of aging. *Neurosci. Biobehav. Rev.* **2008**, *32*, 863–881. [CrossRef] [PubMed]
10. Richoz, A.R.; Lao, J.; Pascalis, O.; Caldara, R. Tracking the recognition of static and dynamic facial expressions of emotion across the life span. *J. Vis.* **2018**, *18*, 5. [CrossRef] [PubMed]
11. Chiang, Y.C.; Chien, S.H.L.; Lyu, J.L.; Chang, C.K. Recognition of dynamic emotional expressions in children and adults and its associations with empathy. *Sensors* **2024**, *24*, 4674. [CrossRef] [PubMed]
12. Ekman, P.; Friesen, W.V. Constants across cultures in the face and emotion. *J. Personal. Soc. Psychol.* **1971**, *17*, 124–129. [CrossRef] [PubMed]
13. Ekman, P.; Friesen, W.V. *Unmasking the Face: A Guide to Recognizing Emotions from Facial Clsues*; Prentice-Hall: Englewood Cliffs, NJ, USA, 1975.
14. Tottenham, N.; Tanaka, J.W.; Leon, A.C.; McCarry, T.; Nurse, M.; Hare, T.A.; Marcus, D.J.; Westerlund, A.; Casey, B.; Nelson, C. The NimStim set of facial expressions: Judgments from untrained research participants. *Psychiatry Res.* **2009**, *168*, 242–249. [CrossRef] [PubMed]
15. Dalrymple, K.A.; Gomez, J.; Duchaine, B. The Dartmouth Database of Children's Faces: Acquisition and Validation of a New Face Stimulus Set. *PLoS ONE* **2013**, *8*, e79131. [CrossRef] [PubMed]

16. Franz, M.; Müller, T.; Hahn, S.; Lundqvist, D.; Rampoldt, D.; Westermann, J.-F.; Nordmann, M.A.; Schäfer, R.; Li, Z. Creation and validation of the Picture-Set of Young Children's Affective Facial Expressions (PSYCAFE). *PLoS ONE* **2021**, *16*, e0260871. [CrossRef] [PubMed]
17. LoBue, V.; Thrasher, C. The Child Affective Facial Expression (CAFE) set: Validity and reliability from untrained adults. *Front. Psychol.* **2015**, *5*, 1532. [CrossRef] [PubMed]
18. Negrão, J.G.; Osorio, A.A.C.; Siciliano, R.F.; Lederman, V.R.G.; Kozasa, E.H.; D'ANtino, M.E.F.; Tamborim, A.; Santos, V.; de Leucas, D.L.B.; Camargo, P.S.; et al. The child emotion facial expression set: A database for emotion recognition in children. *Front. Psychol.* **2021**, *12*, 666245. [CrossRef] [PubMed]
19. Romani-Sponchiado, A.; Sanvicente-Vieira, B.; Mottin, C.; Hertzog-Fonini, D.; Arteche, A. Child Emotions Picture Set (CEPS): Development of a database of children's emotional expressions. *Psychol. Neurosci.* **2015**, *8*, 467–478. [CrossRef]

children

MDPI

Article

Children's Socioemotional Strengths in Early Childhood Education (ECE) and Before/After School Care (BASC): A Multilevel Ecological Analysis

Imogen M. Sloss [1], Nicola Maguire [2] and Dillon T. Browne [1,*]

[1] Department of Psychology, University of Waterloo, 200 University Ave. W., Waterloo, ON N2L 3G1, Canada; isloss@uwaterloo.ca

[2] The Learning Enrichment Foundation, 116 Industry St., Toronto, ON M6M 4L8, Canada

* Correspondence: dillon.browne@uwaterloo.ca

Abstract

Background/Objectives: The current study explored how trajectories of children's socioemotional strengths were explained by school, classroom, and individual differences in the context of licensed early childhood education (ECE), involving preschool and before/after school programming. The predictive role of neighbourhood socioeconomic status (SES) was also explored. **Methods:** Participants included $n = 226$ children from 39 classrooms across seven ECE centres in a large city in Canada. Educators completed measures of children's socioemotional strengths at three time points between January and June 2024. Children's forward sortation areas (FSA) were also linked with publicly available data on neighbourhood SES from the 2021 census. Four-level multilevel models estimated scores across time, individual, classroom, and school levels. **Results:** All four levels significantly explained variance in strengths. On average, child strengths improved over the 4.5 months of ECE programming. Random slopes at the individual and classroom level revealed variability in trajectories. Higher neighbourhood SES was associated with higher socioemotional strengths and was not associated with change over time. **Conclusions:** The findings of this study reveal that child, classroom, school, and neighbourhood factors interact to foster child socioemotional strengths. Thus, targeted and universal programs for promoting socioemotional development in ECE must similarly adopt a multiple levels of analysis perspective.

Keywords: socioemotional strengths; early childhood education; socioeconomic status; neighbourhood census; multilevel modelling

1. Introduction

Licensed early childhood education (ECE) and before-and-after-school care (BASC) are contexts that nurture children's socioemotional strengths [1–5], where investments in early life yield the greatest lifetime return [6]. Licensed ECE and BASC providers are usually required to engage in empirical monitoring exercises, ensuring that children are growing in their socioemotional strengths, supporting accountability, while also promoting quality enhancement [7]. To date, there is a paucity of Canadian literature examining variation in children's socioemotional strengths from a multilevel and ecological perspective, identifying important sources of variation not only across time but in addition to stable individual, classroom, and center-level differences. While such approaches have been applied in the US [1,8–11] and for academic outcomes [12], it remains an understudied question for ECE and BASC, especially within the unique Canadian context [13]. This information can be helpful in guiding policy and program planning investments. Therefore, the current study sought to explore trajectories of socioemotional strengths among children enrolled in licensed ECE (including preschool programs and BASC, as these are both part of ECE in Canada) with a major provider in a large Canadian city. A multilevel and ecological perspective was employed, whereby classroom and school-level differences in trajectories were investigated. Following perspectives highlighting the importance of social determinants of child and family well-being [14], the predictive role of census-linked socioeconomic status of children's neighbourhood of residence was also explored.

1.1. Socioemotional Strengths

Socioemotional strengths begin to develop during infancy, and are associated with educational, occupational, physical, emotional, social, and behavioural outcomes across the lifespan [15,16]. Multiple aspects of children's environments interact to foster or hinder the development of these strengths [1,4,17]. The Collaborative for Academic, Social, and Emotional Learning [18] explains that socioemotional strengths allow individuals to develop and work towards goals, persist in the face of setbacks, understand and regulate emotions, cope effectively with challenges, build secure and healthy relationships, and understand others' perspectives. These strengths are often grouped into the following five key areas: decision making, self-management, self-awareness, social awareness, and relationship skills [18]. Researchers have found that socioemotional strengths play a significant role in myriad long-term outcomes, including life satisfaction, physical and mental health, socioeconomic status (SES), substance use, criminal activity, and social relationships [3,15,16,19]. Thus, exploring variation in socioemotional strengths across layers of the developmental ecosystem can help guide families, educators, service providers, and policy makers to nurture these strengths and their associated outcomes.

Various researchers have outlined the developmental systems that give rise to socioemotional strengths [20–22]. These strengths are the product of a complex interaction between a child's temperament, genetics, and life experiences. Socioemotional strengths begin to develop as an infant forms an attachment with their primary caregiver. Over the first years of life, this growth expands to include the development of emotion regulation, self-control, attention, social skills, empathy, and autonomy [22]. As biological maturation propagates in a supportive social surrounding, the developmental ecology becomes more complex, and children's strengths are further shaped by various contexts in a child's life, including their family, neighbourhood, classroom, care providers, school, and peers. Furthermore, children's experiences are informed by sociodemographic contexts, including SES, which are unequally distributed due to the concentration of resources within various groups in society [14]. Therefore, some groups tend to experience greater economically induced stress and limited access to development-enhancing resources and opportunities [14]. Empirical studies of children's socioemotional strengths should measure and model these influences, given their demonstrable role in predicting outcomes. Additionally, these studies should operationalize socioemotional strengths in accordance with the sample's developmental stage [23]. Therefore, the current study employed valid and developmentally sensitive measures of socioemotional strengths [24–26].

1.2. Contextual Influences on Socioemotional Strengths

1.2.1. Socioeconomic Status (SES)

SES is a latent dimension comprising various economic indicators including an individual, family, or neighbourhood's level of education, income, and access to economic capital (e.g., assets). Developmental researchers often focus on SES' role in children's growth and development, which remains an essential domain of research and policy. Indeed,

https://doi.org/10.3390/children13010023

researchers have found that children from lower-SES families have lower socioemotional strengths compared to children from higher-SES families [4,8,27]. Differences in these strengths are observable before children enter school [28]. Furthermore, the magnitude of this difference in socioemotional strengths gets wider as children get older [4].

There are many reasons why children from lower-SES families experience greater challenges to socioemotional skill development, including access to opportunity and psychosocial stress [28,29]. In general, lower-SES families have access to fewer resources compared to higher-SES families. Limited financial resources mean that lower-SES families may struggle to pay for educational materials and high-quality services or experiences for their children [4], impacting the type of stimulation that children receive. Additionally, neighbourhood factors can impact socioemotional strengths. In some jurisdictions, ECE centres and schools in lower-SES areas receive less funding, reducing the quality of these services [4,30,31]. A study by McCoy and colleagues [31] elucidated how these neighbourhood risk factors may impact development: they found that preschool program quality mediated the relationship between neighbourhood poverty and socioemotional strengths. In addition to difficulties accessing resources, lower-SES families and neighbourhoods tend to experience higher levels of stress and adversity. Masarik & Conger [32] outline how these stressors impact the family system and child well-being. Financial stress and community disadvantage increase caregiver stress, which can spillover into the marital and parent-child relationship [32]. When caregivers experience high levels of stress, they may have less capacity to be supportive and involved in the lives of their children [32]. Furthermore, caregivers experiencing high levels of stress may be less consistent in how they interact with their children [32].

1.2.2. Early Childhood Education and Before/After School Care

Another prominent context in which socioemotional strengths develop is in ECE and BASC. For example, ECE and BASC environments allow children to interact with peers, follow routines, and comply with rules, providing opportunities to foster socioemotional development [1,17]. As a result, many researchers have explored ECE and BASC contexts in relation to socioemotional development. High-quality ECE and BASC programs are associated with many short- and long-term benefits, including socioemotional strengths [1,33]. For example, these high-quality programs are related to well-being, social skills, emotion regulation, inhibitory control, and academic achievement [3,13,34–36]. One pathway in which children develop these strengths is through play, as interacting and playing with peers fosters problem-solving, cooperation, perspective-taking, sharing, and self-restraint [37]. Furthermore, Jennings & Greenberg [21] and Jones & Bouffard [38] present models that outline how the classroom environment can foster socioemotional strengths. They posit that there are bidirectional relationships between educator socioemotional strengths and well-being, student–educator relationships, classroom management, classroom climate, and child socioemotional strengths, which is supported by research [8,39,40]. Moreover, these relationships are influenced by school and community contextual factors, highlighting the importance of considering a child's developmental ecology [4,39,40].

ECE and BASC are demonstrably linked to socioemotional skill development in children. Yet, multilevel and ecological studies of children's developmental health demonstrate systematic variation in outcomes as a function of the multiple and hierarchical levels of a child's developmental system [14]. For example, research by Rasbash and colleagues (2010) suggests that multilevel contexts, including the individual, family, school, and neighbourhood level, each explain variance in children's early educational trajectories, some of which is ostensibly linked to social disadvantage [12]. Notably, ECE appears to counteract some of the negative influence that low-SES environments have on socioemotional skill

https://doi.org/10.3390/children13010023

development, helping children from lower-SES households catch up to their peers [1]. In addition to short-term improvements, ECE programs are related to long-term socioemotional benefits. A meta-analysis of preschool programs for children from low-SES households found that they are associated with positive socioemotional functioning throughout school and into adulthood [10]. Furthermore, Reynolds et al. [11] reported the 19-year outcomes of children who participated in high-quality care versus those who participated in typical care. They found that high-quality care was associated with higher educational attainment and full-time employment, and lower rates of depressive symptoms, criminal activity, disability, and out-of-home placements [11]. At present, it remains unclear how children's socioemotional outcomes in Canadian ECE and BASC vary across the levels of a child's developmental ecology, as studies are often not designed to partition these relative sources of variance. This is problematic, as understanding the multiple sources of variability in socioemotional strengths, and their relative contributions, can help organizations and policy makers direct limited resources to enhancing child well-being.

1.3. The Present Study

The present study aimed to address these gaps by exploring the trajectories of educator-reported socioemotional strengths for children enrolled in ECE centres providing preschool and BASC programs in a Canadian context. The research questions investigated in this study include the following: (1) What proportion of variance in socioemotional strengths is attributed to (a) change over time, (b) child differences, (c) classroom differences, and (d) school differences? (2) How do socioemotional strengths change over 4.5 months of programming? (3) Does neighbourhood-level SES predict socioemotional strengths at baseline and over time?

We hypothesized that all four levels (change, child, classroom, and school) would explain a significant proportion of variance in strengths. Secondly, we predicted that strengths would improve over time. Finally, we expected that children from neighbourhoods with lower SES, as indexed through linkage with the Canadian census, would have lower strengths at baseline, and would improve more during the study period.

2. Materials and Methods

2.1. Participants and Procedure

Study participants were children enrolled in infant (n = 12), toddler (n = 32), preschool (n = 49), and kindergarten/school-aged programming (n = 133) at seven ECE centres run by a community agency in North-West Toronto, Ontario. These centres operate within public school settings and include full-day programs for preschool-aged children, as well as before-and-after school care (BASC) programming for school-aged children, which is a common practice in Canada. Keeping with the best-practice principles of community-based research [41], the agency was empowered to select priorities, including identifying the specific centers that they felt would be optimal in implementing this research protocol. All centres are licensed within the province of Ontario, Canada, and because of the federal government's ECE program, families were able to access subsidies to lower the significant cost of care. The organization emailed caregivers with details about the study, asking if they would like their child to participate. Caregivers provided informed consent to have educators (registered early childhood educators and early childhood assistants) complete strengths-based measures about their children over three time points (January 2024, March 2024, and June 2024). Educators were trained to complete the paper surveys used in the study. They completed the surveys during paid hours when they had classroom coverage from other educators. Upon completion, the paper copies were transferred to the researchers, where research assistants inputted data into Qualtrics. The paper copies were

 https://doi.org/10.3390/children13010023

stored in a secure location at the agency and university. The educators in the study were not considered to be research participants because they were engaging in tasks that were part of their routine job (filling out regular measures for children), and educator information was anonymized; therefore, educators did not need to provide consent. The current convenience sample included n = 226 children and ranged in age from 0.5 to 10 years (M = 4.60, SD = 2.21), which maps onto the age range of children served by the agency partner. Inclusion criteria required that the children were enrolled at one of the seven centres. There were no exclusion criteria for children within these centres. Children were identified as boys (n = 108), girls (n = 108), and gender minorities (n = 10). These children were from n = 39 classrooms in the seven ECE centres. There was variation in the number of children at each time point due to attendance changes (T1, n = 215; T2, n = 213; T3, n = 217), and all children were retained in analyses. Ethics clearance (#40760 and #43229) for this study was obtained by the IRB at [blinded].

2.2. Measures

2.2.1. Outcome Variable: Strengths

School-aged children. The 72-item Devereux Student Strengths Assessment [DESSA; 24] was used to assess educator-reported child strengths for school-aged children (kindergarten through Grade 8, though the oldest children in the study were approximately Grade 4). Educators rated the frequency of all items during the past four weeks on a five-point Likert scale from "Never" to "Very Frequently". Higher scores indicated higher levels of strengths. Sample items include: "offer to help somebody" and "work hard on projects". Internal consistency was excellent (Cronbach's alpha = 0.99).

Preschoolers. Strengths for preschoolers (children aged three to five years) were assessed using the 27-item Devereux Early Childhood Assessment for Preschoolers Second Edition [DECA-P2; 24]. Educators were instructed to rate the frequency of specific behaviours over the past four weeks on a five-point Likert scale from "Never" to "Very Frequently". Higher scores on this measure indicated greater strengths. Example items are: "listen to or respect others" and "keep trying when unsuccessful". This measure had good internal consistency (Cronbach's alpha = 0.85).

Toddlers. The 36-item Devereux Early Childhood Assessment for Toddlers [DECA-T; 26] assessed strengths in children aged 18 to 36 months. Educators rated how frequently (five-point Likert scale from "Never" to "Very Frequently") toddlers engaged in various behaviours. Higher scores reflected higher levels of strengths. Sample items include: "ask to do new things" and "handle frustration well". The internal consistency of this measure was excellent (Cronbach's alpha = 0.97).

Infants. Educators completed the 33-item Devereux Early Childhood Assessment for Infants [DECA-I; 26] to assess the strengths of infants between one to 18 months old. Educators rated the frequency of the items on a five-point Likert scale from "Never" to "Very Frequently", with higher scores indicating higher strengths. Sample items include: "imitate actions of others" and "explore surroundings". This measure had good internal consistency (Cronbach's alpha = 0.92).

Raw scores for all measures of socioemotional strengths were converted to percentiles to contextualize the scores in reference to a normative sample. Although the use of percentile scores can pose limitations in growth models, the present study employed percentiles to allow for the comparison of children at various developmental stages.

2.2.2. Predictor Variable: Neighbourhood Socioeconomic Status

In line with Jones and colleagues [42], neighbourhood-level socioeconomic status (SES) based on a child's Forward Sortation Area (FSA; an alpha-numeric geo-code that

https://doi.org/10.3390/children13010023

describes large neighbourhoods throughout Canada) was taken from the Canadian 2021 Census [43]. An SES composite was created by considering the percentage of adults within an FSA who (1) were unemployed, (2) did not have a high school diploma, (3) were receiving government transfers, (4) led single-parent families, and (5) were living below the low-income line. Factor analysis revealed that these five variables loaded onto one factor, explaining 67.3% of the variance in the data. Higher scores reflected more socioeconomic disadvantage. The internal consistency was good (Cronbach's alpha = 0.87).

2.2.3. Covariates

Covariates included age (in years), gender, and program type. Dummy variables for program type included toddler (toddlers = 1, other = 0), preschooler (preschoolers = 1, other = 0), and school-aged (school-aged = 1, other = 0), which were all compared to infants (0 for all). Dummy variables for gender included girl (girls = 1, boys/gender minorities = 0) and gender minority (gender minorities = 1, girls/boys = 0), who were compared to boys (0 for both).

2.3. *Statistical Analyses*

Four-level multilevel modelling was conducted in RStudio (Version 2025.09.0+387) with the nlme package [44]. Repeated measures were nested in children, who were nested in classrooms, that were nested in schools. The models were compared with the likelihood ratio test and estimated with restricted maximum likelihood. Confidence intervals were created for the fixed and random effects using the profile likelihood method. Outliers for the study variables were winsorized. Assumptions were checked with Q-Q plots for the residuals at all 4 levels. A null model was created for RQ1 to determine the proportion of variance, using intraclass correlation coefficients (ICCs), in strengths due to the four levels of the multilevel model: change over time and error (level 1), child differences (level 2), classroom differences (level 3), and school differences (level 4). To explore the grand mean linear trajectories of strengths over time for RQ2, we added a fixed effect of time (months). Then, we added random intercepts and random slopes at levels 2, 3, and 4, which permitted individual children, classrooms, and schools to have different intercepts and trajectories. The random slope at level 4 was removed in our final model, because of unreasonably large confidence intervals, along with visual inspections of the histogram of residuals, revealing large departures from normality. Lastly, we added covariates and predictor variables as main effects for RQ3, to determine whether they predicted baseline strengths. Predictor variables were also added as an interaction with time to explore whether they predicted change.

Missing Data

Little's MCAR test was conducted for each age group. The results were non-significant for infants ($X^2(219, N = 32) = 192.52$, $p = 0.901$), toddlers ($X^2(777, N = 101) = 736.77$, $p = 0.847$), and preschoolers ($X^2(786, N = 149) = 793.19$, $p = 0.422$). However, Little's MCAR test was significant ($X^2(4082, N = 399) = 4524.36$, $p < 0.001$) for school aged children, revealing that data were not missing at random. To deal with missing data, random forest multiple imputation was conducted with the missForest package [45]. This method uses a single imputation procedure, with evidence to suggest that it outperforms multiple imputation in terms of minimizing parameter bias associated with missingness [46,47].

3. Results

Descriptive statistics and bivariate correlations are presented in Tables 1 and 2.

https://doi.org/10.3390/children13010023

Table 1. Descriptive statistics and bivariate correlations between study variables.

Variable	2	3	4	5	*M*	*SD*
1. Age	−0.21 *	0.11 *	0.08 *	−0.01	4.60	2.21
2. SES disadvantage	–	−0.25 *	−0.14 *	−0.17 *	15.32	4.42
3. Total Strengths T1		–	0.56 *	0.53 *	52.65	27.61
4. Total Strengths T2			–	0.64 *	62.49	26.43
5. Total Strengths T3				–	65.66	27.07

Note. *M* = mean/average; *SD* = standard deviation; SES = socioeconomic status. * denotes significance at the $p < 0.05$ level.

Table 2. Mean strengths by gender, timepoint, and program.

	Boys (0)		Girls (1)	
	M	*SD*	*M*	*SD*
Time 1				
Infants	–*	–*	66.46	19.53
Toddlers	46	33.74	47.75	30.45
Preschoolers	42.81	31.84	52.87	28.56
School	50.38	26.83	60.97	23.11
Time 2				
Infants	–*	–*	76.92	12.49
Toddlers	55.32	36.65	66.15	33.63
Preschoolers	54.38	28.91	63.45	26.32
School	58.11	26.86	67.8	20.68
Time 3				
Infants	–*	–*	–*	–*
Toddlers	64.68	34.11	80.95	19.59
Preschoolers	61.96	35.86	65	23.79
School	56.81	28.91	70.47	18.97

Note. *M* = mean/average; *SD* = standard deviation. Gender minorities were suppressed due to cell size. –* were suppressed due to cell size.

3.1. Question 1: Variance Partitioning

According to the ICCs, each level of the multilevel model, portrayed in Model 1 of Table 3, explained a significant proportion of child strengths in the null model: 32.7% of the variance was due to change (and potentially measurement error), 24.3% was due to child differences, 20.8% was due to classroom differences, and 22.2% was due to school differences.

Table 3. Fixed and random effects of the multilevel model output examining school, classroom, and child trajectories on child socioemotional strengths.

	Model 1	Model 2	Model 3	Model 4
Fixed Effects	*b [95% CI]*	*b [95% CI]*	*b [95% CI]*	*b [95% CI]*
Intercept	54.86 * [44.17, 65.55]	49.54 * [38.75, 60.33]	50.58 * [41.41, 59.76]	67.04 * [50.46, 83.62]
Months		2.69 * [1.95, 3.42]	2.16 * [0.43, 3.90]	2.45 * [1.04, 3.85]
Age				4.04 * [1.56, 6.52]
Toddler				−5.15 [−16.37, 6.07]
Preschool				−24.73 * [−40.28, −9.18]
School				−24.00 * [−42.45, −5.55]
Girl				9.77 * [5.09, 14.46]
Gender minority				5.52 [−5.83, 16.86]
SES disadvantage				−1.38 * [−2.56, −0.19]
Months * SES disadvantage				0.02 [−0.23, 0.27]
Random Effects	*SD [95% CI]*	*SD [95% CI]*	*SD [95% CI]*	*SD [95% CI]*

https://doi.org/10.3390/children13010023

Table 3. *Cont.*

	Model 1	Model 2	Model 3	Model 4
Level 4				
Random intercept	12.77 * [6.00, 27.18]	12.77 * [6.00, 27.18]	10.37 * [4.39, 24.51]	10.32 * [4.36, 24.41]
Random slope			1.44 * [0.37, 5.65]	
Cov (Intercept/Months)			0.79 [−0.95, 1.00]	
Level 3				
Random intercept	11.97 * [8.41, 17.04]	11.97 * [8.41, 17.03]	11.03 * [7.21, 16.89]	10.40 * [6.64, 16.28]
Random slope			3.54 * [2.57, 4.88]	3.68 * [2.71, 4.99]
Cov (Intercept/Months)			−0.20 [−0.60, 0.28]	−0.23 [−0.65, 0.30]
Level 2				
Random intercept	13.95 * [11.81, 16.47]	14.37 * [12.30, 16.79]	17.98 * [15.50, 20.86]	17.13 * [14.69, 19.98]
Random slope			1.39 * [0.83, 2.33]	1.57 * [0.98, 2.52]
Cov (Intercept/Months)			−0.98 [−1.00, 0.31]	−0.98 * [−1.00, −0.25]
Level 1				
Random intercept	18.80 * [17.62, 20.07]	17.82 * [16.70, 19.03]	15.36 * [14.35, 16.44]	15.30 * [14.29, 16.38]

Note. b = unstandardized estimates; SD = standard deviation; *95% CI* = 95% confidence intervals. * denotes significant findings based on the *95% CIs*.

3.2. Question 2: Growth Model

The growth model for child strengths is presented in Figure 1 and Models 2 and 3 in Table 3. On average, strengths improved over time (b = 2.69, 95% CI = 1.95, 3.42), improving approximately 2.69 percentile points each month. Significant random slopes (displayed in Figure 2) at level 2 (child level; b = 1.39, 95% CI = 0.83, 2.33), level 3 (classroom level; b = 3.54, 95% CI = 2.57, 4.88), and level 4 (school level; b = 1.44, 95% CI = 0.37, 5.65), revealed that individual children, classrooms, and schools had different slopes. In Model 3, the covariance was not significant at any level, suggesting that starting values of strengths did not affect rate of change. However, in Model 4, the child-level covariance was significant: children who started with lower strengths improved more over time compared to children with higher initial strengths

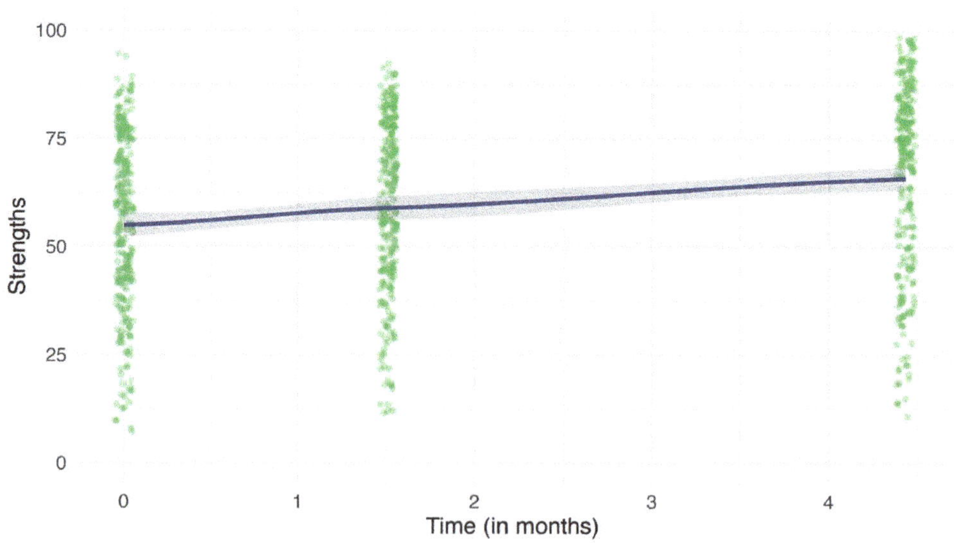

Figure 1. Trajectories of child strengths. The green dots represent individual participant values at each time. The blue line represents the mean trajectory for all participants over time. The grey shading represents *95% CIs*.

https://doi.org/10.3390/children13010023

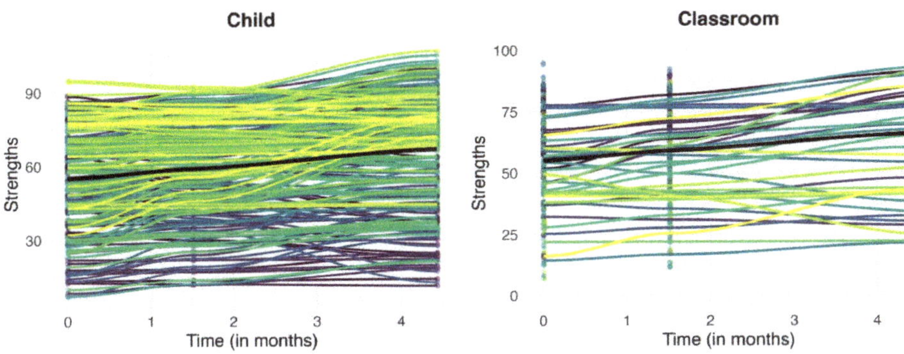

Figure 2. Child- & classroom-level trajectories of child strengths. Each coloured line represents the trajectories of an individual child (**graph on left**) and an individual classroom (**graph on right**). The black line represents the mean trajectory for all participants.

3.3. Question 3: Predictor Variables

Various predictor variables, displayed in Model 4 of Table 3, were significantly associated with child strengths at baseline and over time. Older age was associated with higher baseline strengths (b = 4.04, *95% CI* = 1.56, 6.52). Gender also significantly predicted strengths: girls had significantly higher strengths than boys at baseline (b = 9.77, *95% CI* = 5.09, 14.46). Gender minorities did not significantly differ from girls or boys. There were also significant differences in strengths by program: infants and toddlers had significantly higher strengths than preschoolers and school-aged children. Infants and toddlers, and preschoolers and school-aged children did not significantly differ from one another in strengths. Neighbourhood SES predicted strengths at baseline (b = −1.38, *95% CI* = −2.56, −0.19) and was not associated with change over time (see Figure 3). In other words, greater levels of socioeconomic disadvantage corresponded to lower socioemotional strengths, and this pattern was relatively stable over time. Stated differently, all children improved in their strengths equally, irrespective of the socioeconomic status of their neighbourhood. Neighbourhood differences are presented in the Supplementary Materials.

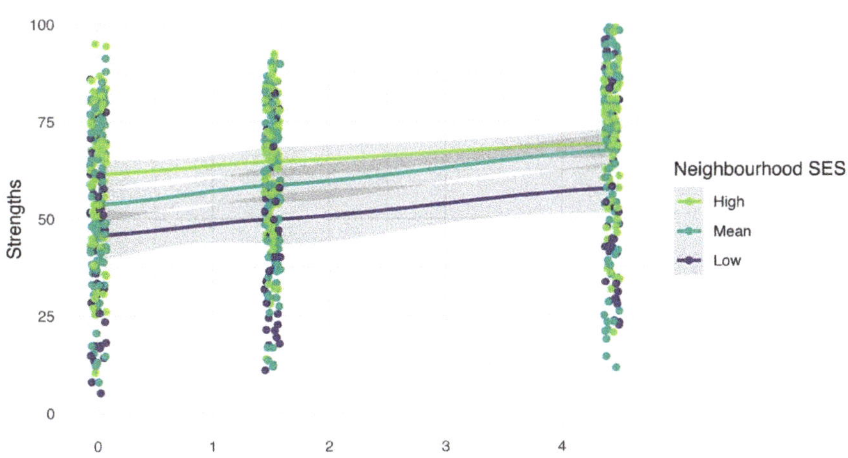

Figure 3. Average child strengths over time by neighbourhood socioeconomic status. Coloured lines represent strength trajectories based on levels of neighbourhood socioeconomic status (SES). The grey shading represents *95% CIs*.

https://doi.org/10.3390/children13010023

3.4. Sensitivity Analysis

To explore the role of developmental stage on child strength trajectories, interactions between program type (infant, toddler, preschool, school-aged) and months were added. None of these interactions were significant, suggesting that, on average, children improved at the same rate regardless of program type.

4. Discussion

The present study sought to investigate the trajectories of socioemotional strengths over 4.5 months of ECE or BASC enrollment, as well as the proportion of variance in strengths attributable to change, individuals, classrooms, and schools in a Canadian context. This research aimed to empirically monitor child trajectories to support accountability and promote quality enhancement for these centres [7]. Furthermore, children's neighbourhood SES was included as an ecological predictor of baseline strengths and change. Our first hypothesis was supported: a significant proportion of socioemotional strengths was explained by all four levels, with change over time (plus measurement error) explaining the highest proportion of variance (32.7%; level 1), followed by child differences (24.3%; level 2), school differences (22.2%; level 4), and classroom differences (20.8%; level 3). Secondly, as expected, socioemotional strengths improved over time. In contrast to our hypothesis, neighbourhood SES predicted strengths at baseline, with children from higher-SES FSAs being reported to have higher strengths. However, neighbourhood SES was not related to change.

4.1. Predictors of Levels and Trajectories of Socioemotional Strengths

Change over time explained the greatest proportion of variance in socioemotional strengths, and linear trajectories suggested that child strengths improved over time, on average. Although our study did not employ a randomized control design, preventing us from concluding causation, it is possible that the ECE/BASC programs may have contributed to these observed improvements. This hypothesis would align with highly controlled research suggesting that high-quality ECE programs are associated with socioemotional strengths, in comparison to programs of lower quality educational supports [1]. Another possible mechanism of change is that as children aged, they developed more socioemotional strengths [22]. While measurements were age-normed and age was a control variable, the present study's longitudinal design meant that time was elapsing while children completed the same set of tools (and norms), meaning that maturation effects are possible. In other words, the positive slope could be related to typical development, rather than the influence of ECE programming itself.

In the present study, the neighbourhood-level SES of children's place of residence was associated with strengths, with children from lower-SES areas being reported to have lower baseline strengths. This finding is consistent with previous research, whereby low-SES is a risk factor for lower levels of socioemotional strengths [4,8,27]. Various mechanisms could explain why low-SES contexts may act as a risk factor in socioemotional development. Firstly, lower-SES areas tend to have fewer community resources due to levels of funding [4]. As a result, children may have fewer opportunities to access important services, and services that they do access may be less promotive of positive development [4]. Additionally, caregivers in these homes are at a greater likelihood of experiencing higher levels of financial- and community-related stressors, which could place strain on mental health and family functioning [4,32]. Consequently, caregivers may be less able to provide high levels of monitoring and consistency, both of which are related to socioemotional development [27]. Lastly, children from lower-SES neighbourhoods tend to be exposed to higher levels of adversity within the community v [4]. In response to these conditions,

https://doi.org/10.3390/children13010023

caregivers may be hesitant to allow their children to play outside, potentially reducing physical activity, an important predictor of socioemotional development [13,48,49].

Unfortunately, research suggests that the gap in socioemotional strengths between children from higher versus lower SES contexts widens as children get older [4]. However, in the current study, we found that, on average, all children improved at the same rate, regardless of neighbourhood SES. Although the children from lower-SES areas did not catch up to their peers during the study period–a replicated finding from previous ECE research [1,10,11]–it is possible that ECE programming prevented this gap from widening. Furthermore, many studies, including a 19-year longitudinal study by Reynolds and colleagues [11], explored long-term outcomes following ECE [1,10]. It is possible that our study was too short to capture the protective role of high-quality ECE for children from lower-SES areas. Nevertheless, our study was not a randomized control trial, preventing us from disentangling the impact of ECE versus maturation and other contexts in the children's lives.

Although child strengths tended to improve over time, the significant random effects at the child, classroom, and school level revealed that children, classrooms, and schools displayed unique trajectories. Factors at these three levels of analysis will be explored below.

4.2. Socioemotional Strengths as an Individual-Level Phenomenon

The individual level explained the second largest proportion of variance in socioemotional strengths, suggesting that children systematically differ from one another in terms of strengths. Child age and gender were two individual-level variables that were found to predict socioemotional strengths. In the present study, age was positively associated with socioemotional strengths, suggesting that strengths increase as children get older, which replicates the findings of Hu et al. [8]. As mentioned above, children develop greater socioemotional strengths as they age, potentially contributing to this observed improvement. On average, girls had higher strengths compared to boys, which is consistent with previous research on socioemotional strengths [50]. Girls tend to be encouraged to express emotions, and reinforced for sensitive, compassionate and prosocial behaviour [51]. In contrast, boys may be encouraged to hide emotions other than anger, and to act in a more dominant and assertive way [51]. Given this approach to socialization, the American Psychological Association [52] has reported that exploring socioemotional development in boys and men is an important area for future inquiry. Family factors, which were not explored or isolated in the current study, may have also contributed to the child-level variance in strengths. Previous research has found that childhood adversity, parenting, and family functioning are associated with variation in socioemotional strengths [1,4,8,27]. Future studies might consider these family-level predictors of socioemotional strengths, given the substantial proportion of variance explained by the between-child level.

4.3. Socioemotional Strengths as a Classroom-Level Phenomenon

Jennings & Greenberg [21] and Jones & Bouffard [38] discuss the importance of teachers and schools in promoting socioemotional development. They assert that there are bidirectional relationships between a teacher's own well-being and socioemotional strengths, student–teacher relationships, classroom management, classroom climate, socioemotional learning curriculum implementation, and student socioemotional strengths. Furthermore, they discuss how all these processes occur in the context of the school and community, highlighting the importance of considering the classroom ecology. In the current model, the classroom level explained the smallest—although still significant and sizable—proportion of variance in socioemotional strengths. This finding aligns with previous multilevel research in the context of ECE [8,9], including research in a Canadian context [13], and with theo-

https://doi.org/10.3390/children13010023

retical models [21,38]. Unmeasured classroom- and educator-related variables may have contributed to this variance. Classroom organization, structure, and management are associated with socioemotional strengths [8,40]. Teaching techniques that involve small-group collaborative activities are associated with the development of socioemotional strengths [39]. In addition, classroom environments that foster a growth mindset are conducive to socioemotional development. For example, educators' use of language (e.g., moving from "you can't do it" to "you can't do it yet") and teaching strategies (e.g., providing students with the opportunity to redo work) can foster socioemotional strengths, such as positive coping, persistence, and goal-directed behaviour [39]. Student–educator relationships also play a role in positive socioemotional outcomes [39,40]; therefore, educator turnover negatively impacts socioemotional development [39]. Lastly, educators who have higher socioemotional strengths, and who experience lower levels of stress and mental health difficulties tend to be more sensitive to children's mental and emotional states, and, therefore, have a greater ability to foster these strengths in students [39,40,53]. Educators should strive to implement these classroom management strategies and teaching styles, as well as personal self-care strategies, to ensure that their classroom has the potential to foster socioemotional strengths.

4.4. *Socioemotional Strengths as a School-Level Phenomenon*

The results revealed that the school level explained a significant proportion of variance in child strengths. These findings differ somewhat from previous multilevel modelling studies, whereby the school level did not significantly contribute to variance in socioemotional strengths in ECE settings [8,9]. However, when exploring other outcomes, such as academic achievement, multilevel modelling studies have identified the school level as a significant source of variability [54,55]. Although not explored in the current study, school characteristics may contribute to the variance observed at this level. School funding has implications for student socioemotional development. For example, schools with more funding tend to have more educational resources, lower student–educator ratios, and the ability to hire the most experienced educators [4]. Furthermore, counselling and emotional support services are sometimes available to support youth, fostering their socioemotional strengths [39]. In addition, schools with more funding may be able to provide socioemotional strengths training, explicitly equipping educators to foster these strengths [39,40]. School culture can also play a role in student strengths. For example, schools that are committed to fostering equity and inclusivity may implement programs that support students from underrepresented groups [39]. Furthermore, support between staff members is key to a school climate that fosters socioemotional development [39,40]. Lastly, relational discipline strategies that support students in repairing harm and learning from their behaviour can promote socioemotional strengths [39]. Future Canadian ECE research that explicitly measures these school-level (and classroom-level) factors will be informative for service providers, organizations, and policy makers in planning programming and educator development aimed at growing children's socioemotional well-being.

4.5. *Implications*

There are several implications that emerge from this multilevel study of children's ECE and BASC. First, the comparable rate of growth in socioemotional strengths between children from lower- and higher-SES communities suggests that ECE and BASC programs might promote socioemotional development among families who are facing economic uncertainty. These findings highlight the importance of expanding low-cost, high-quality ECE programs for lower-SES families, and are informative considering Canadian national programs to make ECE and BASC more affordable for families. However, future longer-term studies should be conducted to further explore the relationship between SES and

https://doi.org/10.3390/children13010023

strengths in a Canadian context. The result that the highest proportion of variance was explained by change over time suggests that socioemotional strengths do change during infancy, toddler years, and childhood. Therefore, families, educators, and policy makers should capitalize on this formative period by seeking to promote environmental predictors of child strengths at the individual, classroom, school, and neighbourhood level. The Heckman Equation (n.d.) provides recommendations for fostering positive lifelong skills. Supporting socioemotional development should begin at birth, and ECE programs for infants as young as eight weeks old are associated with greater school readiness. Furthermore, they emphasize the importance of continuous care, as this provides consistency in relationships, a key component in the formation of secure attachments. In addition, they assert the critical role that parents play in child development, and advise that ECE programs involve parents (e.g., psychoeducation), given that the family environment plays a significant role in socioemotional strengths. Next, they recommend that educators and teachers receive training to develop their own socioemotional strengths and learn how to interact with students in ways that promote positive development. Lastly, the Heckman Equation (n.d.) emphasizes the importance of considering skills across domains (e.g., socioemotional strengths, cognitive ability, academic skills), as these areas are interrelated, and influence each other throughout development. ECE providers should strive to implement these recommendations, given the significant role that the ECE context plays in child development.

4.6. Strengths, Limitations, and Future Directions

The present study had several strengths. The proportion of variance explained by multilevel contexts could be identified due to the hierarchical and longitudinal nature of the study data. Additionally, this research adds to the literature about ECE and BASC in a Canadian setting. Another key strength of this study is that information from the Canadian census could be linked to individual participants, allowing for the exploration of neighbourhood-level SES disadvantage on trajectories of strengths.

The current study also had various limitations. Firstly, all measures were completed by educators, which may have inflated the classroom-level variance. In addition, educator ratings may not apply to a child's strengths in contexts outside of ECE. Future studies could include ratings from multiple informants or an observational design. Future research would also benefit from classroom- and school-level predictor variables. Next, our predictor variables explored SES at the neighbourhood level as opposed to the family or household level. While neighbourhood SES that is directly linked from the census is a robust and commonly used metric, it is also true that there is within-neighbourhood variation in SES as well. Therefore, a strength of the current study is that we were able to explore the role of a child's neighbourhood in their socioemotional development. However, we did not have detailed family-level information, which may have further clarified the relationship between SES and socioemotional strengths. Additionally, while we explored the proportion of variance explained by school and classroom differences, we did not include any predictors at these levels, preventing us from understanding how these environments affect socioemotional strengths. Furthermore, the measures assessed different strengths across developmental stages, complicating the interpretability of results; however, the moderation analysis aimed to explore developmental differences in trajectories (see sensitivity analyses). Next, our use of percentiles instead of raw scores in our growth models was a limitation, as the trajectory of the statistically average child would be flat. Yet, the proximity of repeated measures allowed us to isolate and identify growth in strengths within a relatively short developmental period (i.e., less than half a year). In addition, all children were enrolled at various centres run by the same organization, which may limit

https://doi.org/10.3390/children13010023

generalizability beyond this organization. Selection bias may also have influenced the results, as there may have been systematic differences between families who did and did not consent to participate in the study. Lastly, we cannot conclude whether ECE participation caused the observed improvements in strengths because the current study did not employ a randomized controlled trial design. Future studies should include random assignment and a control group to disentangle the influence of ECE from other multilevel predictors of socioemotional development.

5. Conclusions

The current study aimed to investigate trajectories of children's socioemotional strengths in the context of ECE, along with the role that contextual factors across multiple levels of analysis play in the development of these strengths. The results revealed that a significant proportion of variance in strengths was attributed to school, classroom, child, and time differences (RQ1). On average, children improved in socioemotional strengths over the 4.5 months; although, some schools, classrooms, and children improved at a faster rate than others (RQ2). Lastly, lower neighbourhood SES was associated with lower strengths at baseline but was not associated with change over time (RQ3). The findings from the current study should be interpreted with caution, as all participants were from centres within the same organization. The findings of the current study emphasize the importance of comprehensively considering a child's multilevel context to foster positive socioemotional development. The relatively equal weight across temporal, individual, classroom, and school differences speak to the balancing of universal and targeted strategies when designing programming and policy aimed at promoting children's socioemotional strengths.

Supplementary Materials: The following supporting information can be downloaded at: https://www.mdpi.com/article/10.3390/children13010023/s1, Figure S1: Total strengths for children enrolled in early childhood education (ECE) between January and April 2024, and census-linked socioeconomic disadvantage by neighbourhood. *Caption*: Neighbourhoods are defined by forward sortation area (FSA), which are the first three characters of a postal code. SES disadvantage was converted to *z*-scores.

Author Contributions: Conceptualization, I.M.S., N.M. and D.T.B.; Methodology, I.M.S., N.M. and D.T.B.; Formal analysis, I.M.S.; Resources, N.M.; Data curation, I.M.S.; Writing—original draft, I.M.S.; Writing—review & editing, I.M.S. and D.T.B.; Visualization, I.M.S.; Supervision, D.T.B.; Project administration, I.M.S. and N.M.; Funding acquisition, D.T.B. All authors have read and agreed to the published version of the manuscript.

Funding: This work was supported by the Canada Research Chairs [950-232347, 2019-2024], the Ontario Early Researcher Award [ER19-15-224, 2021-2026], and the Learning Enrichment Foundation.

Institutional Review Board Statement: The study was approved by the Research Ethics Board of the University of Waterloo (protocol number: #40760; approval date: 10 May 2019).

Informed Consent Statement: Informed consent was obtained from all subjects involved in the study.

Data Availability Statement: The dataset is not available to the public.

Conflicts of Interest: The authors declare no conflicts of interest.

References

1. Gomez, R. Sustaining the benefits of early childhood education experiences: A research overview. *Voices Urban Educ.* **2016**, *43*, 5–14.
2. Minney, D.; Garcia, J.; Altobelli, J.; Perez-Brena, N. Social-emotional learning and evaluation in after-school care: A working model. *J. Youth Dev.* **2019**, *14*, 130–145. [CrossRef]

3. Mondi, C.F.; Giovanelli, A.; Reynolds, A.J. Fostering socio-emotional learning through early childhood intervention. *Int. J. Child Care Educ. Policy* **2021**, *15*, 6. [CrossRef]
4. McLoyd, V.C. Socioeconomic disadvantage and child development. *Am. Psychol.* **1998**, *53*, 185–204. [CrossRef]
5. Wade, C.E. The longitudinal effects of after-school program experiences, quantity, and regulatable features on children's social–emotional development. *Child. Youth Serv. Rev.* **2015**, *48*, 70–79. [CrossRef]
6. Heckman Equation. ABC/CARE: Elements of Quality Early Childhood Programs that Produce Quality Outcomes. 2025. Available online: https://heckmanequation.org/resource/abc-care-elements-of-quality-early-childhood-programs-that-produce-quality-outcomes/ (accessed on 15 May 2025).
7. Roberts-Holmes, G.; Bradbury, A. The datafication of early years education and its impact upon pedagogy. *Improv. Sch.* **2016**, *19*, 119–128. [CrossRef]
8. Hu, B.Y.; Li, Y.; Wang, C.; Wu, H.; Vitiello, G. Preschool teachers' self-efficacy, classroom process quality, and children's social skills: A multilevel mediation analysis. *Early Child. Res. Q.* **2021**, *55*, 242–251. [CrossRef]
9. Kiviruusu, O.; Björklund, K.; Koskinen, H.L.; Liski, A.; Lindblom, J.; Kuoppamäki, H.; Alasuvanto, P.; Ojala, T.; Sam-posalo, H.; Harmes, N.; et al. Short-term effects of the "Together at School" intervention program on children's socio-emotional skills: A cluster randomized controlled trial. *BMC Psychol.* **2016**, *4*, 27. [CrossRef]
10. Nelson, G.; Westhues, A.; MacLeod, J. A meta-analysis of longitudinal research on preschool prevention pro-grams for children. *Prev. Treat.* **2003**, *6*, 31a. [CrossRef]
11. Reynolds, A.J.; Temple, J.A.; Ou, S.; Robertson, D.L.; Mersky, J.P.; Topitzes, J.W.; Niles, M.D. Effects of a school-based, early childhood intervention on adult health and well-being: A 19-year follow-up of low-income families. *Arch. Pediatr. Adolesc. Med.* **2007**, *161*, 730–739. [CrossRef]
12. Rasbash, J.; Leckie, G.; Pillinger, R.; Jenkins, J. Children's educational progress: Partitioning family, school and area effects. *J. R. Stat. Soc. Ser. A* **2010**, *173*, 657–682. [CrossRef]
13. Zhang, J.; Sloss, I.M.; Maguire, N.; Browne, D.T. Physical activity and social-emotional learning in Canadian children: Multilevel perspectives within an early childhood education and care setting. *Soc. Emot. Learn. Res. Pract. Policy* **2024**, *4*, 100069. [CrossRef]
14. Browne, D.T.; McArthur, B.A.; Racine, N. Editorial: Multilevel social determinants of individual and family well-being: National and international perspectives. *Front. Epidemiol.* **2024**, *4*, 1381516. [CrossRef]
15. Danner, D.; Lechner, C.M.; Spengler, M. Editorial: Do we need socio-emotional skills? *Front. Psychol.* **2021**, *12*, 723470. [CrossRef] [PubMed]
16. Jones, D.E.; Greenberg, M.; Crowley, M. Early social-emotional functioning and public health: The relationship between kindergarten social competence and future wellness. *Am. J. Public Health* **2015**, *105*, 2283–2290. [CrossRef]
17. Fisher, A.E.; Johnson, L.R.; Minnes, S.; Miller, E.K.; Riccardi, J.S.; Dimitropoulos, A. Predictors of social emotional learning in after-school programming: The impact of relationships, belonging, and program engagement. *Psychol. Sch.* **2024**, *61*, 1318–1335. [CrossRef]
18. CASEL. What Is the CASEL Framework? Available online: https://casel.org/fundamentals-of-sel/what-is-the-casel-framework/ (accessed on 15 May 2025).
19. Kim, E.K.; Allen, J.P.; Jimerson, S.R. Supporting student social emotional learning and development. *Sch. Psychol. Rev.* **2024**, *53*, 201–207. [CrossRef]
20. Eisenberg, D.; Hunt, J.; Speer, N. Mental health in American colleges and universities: Variation across student subgroups and across campuses. *J. Nerv. Ment. Dis.* **2013**, *201*, 60–67. [CrossRef] [PubMed]
21. Jennings, P.A.; Greenberg, M.T. The prosocial classroom: Teacher social and emotional competence in relation to student and classroom outcomes. *Rev. Educ. Res.* **2009**, *79*, 491–525. [CrossRef]
22. Malik, F.; Marwaha, R. *Developmental Stages of Social Emotional Development in Children*; StatPearls Publishing: Treasure Island, FL, USA, 2022.
23. Halle, T.G.; Darling-Churchill, K.E. Review of measures of social and emotional development. *J. Appl. Dev. Psychol.* **2016**, *45*, 8–18. [CrossRef]
24. LeBuffe, P.A.; Naglieri, J.A. *Devereux Early Childhood Assessment for Preschoolers*, 2nd ed.; Kaplan Early Learning Company: Lewisville, NC, USA, 2012.
25. LeBuffe, P.A.; Shapiro, V.B.; Naglieri, J.A. *Devereux Student Strengths Assessment: K-8th Grade: A Measure of Social-Emotional Competencies of Children in Kindergartern Through Eight Grade*; Aperture Education: Fort Mill, SC, USA, 2014. Available online: https://measuringsel.casel.org/wp-content/uploads/2018/10/DESSA-User-Manual.pdf (accessed on 15 May 2025).
26. Mackrain, M.; LeBuffe, P.; Powell, G. *Devereux Early Childhood Assessment for Infants and Toddlers*; Kaplan Early Learning Company: Lewisville, NC, USA, 2007.
27. Repetti, R.L.; Taylor, S.E.; Seeman, T.E. Risky families: Family social environments and the mental and physical health of offspring. *Psychol. Bull.* **2002**, *128*, 330–366. [CrossRef] [PubMed]

https://doi.org/10.3390/children13010023

28. Browne, D.T.; Wade, M.; Prime, H.; Jenkins, J.M. School readiness amongst urban Canadian families: Risk profiles and family mediation. *J. Educ. Psychol.* **2018**, *110*, 133–146. [CrossRef]
29. Bradley, R.H.; Corwyn, R.F. Socioeconomic status and child development. *Annu. Rev. Psychol.* **2002**, *53*, 371–399. [CrossRef] [PubMed]
30. Adamson, F.; Darling-Hammond, L. Funding disparities and the inequitable distribution of teachers: Evaluating sources and solutions. *Educ. Policy Anal. Arch.* **2012**, *20*, 37. [CrossRef]
31. McCoy, D.C.; Connors, M.C.; Morris, P.A.; Yoshikawa, H.; Friedman-Krauss, A.H. Neighborhood economic disadvantage and children's cognitive and social-emotional development: Exploring Head Start classroom quality as a mediating mechanism. *Early Child. Res. Q.* **2015**, *32*, 150–159. [CrossRef]
32. Masarik, A.S.; Conger, R.D. Stress and child development: A review of the Family Stress Model. *Curr. Opin. Psychol.* **2017**, *13*, 85–90. [CrossRef]
33. Murano, D.; Sawyer, J.E.; Lipnevich, A.A. A meta-analytic review of preschool social and emotional learning interventions. *Rev. Educ. Res.* **2020**, *90*, 227–263. [CrossRef]
34. Blewitt, C.; Fuller-Tyszkiewicz, M.; Nolan, A.; Bergmeier, H.; Vicary, D.; Huang, T.; McCabe, P.; McKay, T.; Skouteris, H. Social and emotional learning associated with universal curriculum-based interventions in early childhood education and care centers: A systematic review and meta-analysis. *JAMA Netw. Open* **2018**, *1*, e185727. [CrossRef]
35. Durlak, J.A.; Weissberg, R.P.; Pachan, M. A meta-analysis of after-school programs that seek to promote personal and social skills in children and adolescents. *Am. J. Community Psychol.* **2010**, *45*, 294–309. [CrossRef]
36. Luo, L.; Reichow, B.; Snyder, P.; Harrington, J.; Polignano, J. Systematic review and meta-analysis of classroom-wide social–emotional interventions for preschool children. *Top. Early Child. Spec. Educ.* **2022**, *42*, 4–19. [CrossRef]
37. Ashiabi, G.S. Play in the preschool classroom: Its socioemotional significance and the teacher's role in play. *Early Child. Educ. J.* **2007**, *35*, 199–207. [CrossRef]
38. Jones, S.M.; Bouffard, S.M. Social and emotional learning in schools: From programs to strategies. *Soc. Policy Rep.* **2012**, *25*, 1–22. [CrossRef]
39. Allbright, T.N.; Marsh, J.A.; Kennedy, K.E. Social-emotional learning practices: Insights from outlier schools. *J. Res. Innov. Teach. Learn.* **2019**, *12*, 35–52. [CrossRef]
40. Schonert-Reichl, K.A. Social and emotional learning and teachers. *Future Child.* **2017**, *27*, 137–155. [CrossRef]
41. Burr, T.; Degotardi, S. Design-based research in early childhood education: A scoping review of methodologies. *Early Child. Educ. J.* **2024**, *53*, 2795–2810. [CrossRef]
42. Jones, L.; Stabile, M.; Koebel, K.; Furzer, J. The effect of household earnings on child school mental health designations: Evidence from administrative data. *J. Hum. Resour.* **2024**, *59*, S41–S76. [CrossRef]
43. Statistics Canada. Census Profile, 2021 Census of Population. 2023. Available online: https://www12.statcan.gc.ca/census-recensement/2021/dp-pd/prof/index.cfm?Lang=E (accessed on 15 May 2025).
44. Pinheiro, J.; Bates, D.; DebRoy, S.; Sarkar, D.; Heisterkamp, S.; Van Willigen, B.; Ranke, J. nlme: Linear and Nonlinear Mixed Effects Models. 2023. Available online: https://cran.r-project.org/web/packages/nlme/index.html (accessed on 15 May 2025).
45. Stekhoven, D.J.; Bühlmann, P. MissForest: Nonparametric missing value imputation using random forest. *Bioinformatics* **2012**, *28*, 112–118. [CrossRef]
46. Graham, J.W. Missing data analysis: Making it work in the real world. *Annu. Rev. Psychol.* **2009**, *60*, 549–576. [CrossRef]
47. Jaafar, N.N.; Rosdi, M.N.A.; Jamaludin, K.R.; Ramlie, F.; Talib, H.A. Imputation analysis of time-series data using a random forest algorithm. In *Intelligent Manufacturing and Mechatronics: iM3F 2023*; Isa, M., Khairuddin, W.H., Razman, I.M.M., Saruchi, M.A., Teh, S., Liu, S.H., Eds.; Lecture Notes in Networks and Systems; Springer: Berlin/Heidelberg, Germany, 2024; Volume 850. [CrossRef]
48. Drenowatz, C.; Eisenmann, J.C.; Pfeiffer, K.A.; Welk, G.; Heelan, K.; Gentile, D.; Walsh, D. Influence of so-cio-economic status on habitual physical activity and sedentary behavior in 8- to 11-year old children. *BMC Public Health* **2010**, *10*, 214. [CrossRef]
49. Wijtzes, A.I.; Jansen, W.; Bouthoorn, S.H.; Pot, N.; Hofman, A.; Jaddoe, V.W.; Raat, H. Social inequalities in young children's sports participation and outdoor play. *Int. J. Behav. Nutr. Phys. Act.* **2014**, *11*, 155. [CrossRef]
50. Fung, W.; Chung, K.K.; Cheng, R.W. Gender differences in social mastery motivation and its rela-tionships to vocabulary knowledge, behavioral self-regulation, and socioemotional skills. *Early Educ. Dev.* **2019**, *30*, 280–293. [CrossRef]
51. Wienclaw, R. Gender Roles. In *Sociology Reference Guide: Gender Roles & Equality*; The Editors of Salem Press, Eds.; Salem Press: Pasadena, CA, USA, 2011; pp. 33–40.
52. American Psychological Association. APA Guidelines for Psychological Practice with Boys and Men. 2018. Available online: http://www.apa.org/about/policy/psychological-practice-boys-men-guidelines.pdf (accessed on 15 May 2025).
53. Parker, S.; Perlman, M.; Prime, H.; Jenkins, J. Caregiver cognitive sensitivity: Measure development and vali-dation in Early Childhood Education and Care (ECEC) settings. *Early Child. Res. Q.* **2018**, *45*, 45–57. [CrossRef]

 https://doi.org/10.3390/children13010023

54. Ma, X.; Klinger, D.A. Hierarchical linear modeling of student and school effects on academic achievement. *Can. J. Educ.* **2000**, *25*, 41–55. [CrossRef]
55. McCoach, D.B.; Gubbins, E.J.; Foreman, J.; Rubenstein, L.D.; Rambo-Hernandez, K.E. Evaluating the Efficacy of Using Predifferentiated and Enriched Mathematics Curricula for Grade 3 Students: A Multisite Cluster-Randomized Trial: A Multisite Cluster-Randomized Trial. *Gift. Child Q.* **2014**, *58*, 272–286. [CrossRef]

 https://doi.org/10.3390/children13010023

children

MDPI

Article

Reducing Preschool Exclusionary Discipline Practices Through Infant and Early Childhood Mental Health Consultation: Findings from the Jump Start Program

Yaray Agosto [1], Morgan D. Darabi [1], Ana Robleto [1], Maite Schenker [1], Bianca Caceres [2], Elizabeth Erban [3], Tania Ramirez [4], Rachel Spector [5] and Ruby Natale [1,*]

[1] Mailman Center for Child Development, University of Miami Miller School of Medicine, Miami, FL 33136, USA; yagosto@med.miami.edu (Y.A.); mdarabi@miami.edu (M.D.D.); arobleto@med.miami.edu (A.R.); mxs2515@med.miami.edu (M.S.)

[2] Community Health of South Florida, Inc., Miami, FL 33190, USA; bcaceres@chisouthfl.org

[3] Family Central, Inc., Miami Shores, FL 33138, USA; elizabetherban@familycentral.org

[4] Clinical Services Department, Jewish Community Services of South Florida, Miami, FL 33181, USA; tramirez@jcsfl.org

[5] The Children's Trust of Miami-Dade County, Miami, FL 33129, USA; rachel@thechildrenstrust.org

* Correspondence: rnatale@med.miami.edu

Highlights

What are the main findings?

- Childcare centers reported significant reductions in both traditional suspensions/ expulsions and soft expulsions after participation in Jump Start, an infant and early childhood mental health consultation model.
- Jump Start was associated with reductions in soft expulsions when implementation significantly improved childcare center discipline policies.

What are the implications of the main findings?

- Given reductions in traditional suspensions/expulsions and negative developmental trajectories associated with exclusionary discipline practices, researchers, educators, and policy makers should increasingly focus their efforts on the reduction in soft expulsions.
- When aiming to reduce soft expulsions, educators and policy makers should select interventions that support childcare centers in improving their policies related to exclusionary discipline practices.

Abstract

Background/Objectives: National data show that about 250 preschoolers are suspended or expelled daily in the United States. Jump Start is a multi-tiered infant and early childhood mental health consultation program that strengthens early care and education centers' capacity to support children's social–emotional development and prevent school suspension and expulsion. This retrospective study examined center-level exclusionary discipline practices, Jump Start participation, and related changes in discipline and expulsion policies. Methods: Data from 270 early care and education centers across Miami-Dade County that received Jump Start services during one of three academic years (2022–2023, 2023–2024, or 2024–2025) were included. Analyses examined associations between baseline exclusionary discipline practices, program duration, discipline and expulsion policy changes, and post-Jump Start exclusionary discipline practices. Results: Statistically significant reductions were observed in the frequency of traditional suspensions/expulsions and soft expulsions following Jump Start participation. The association between the Jump Start duration and post-Jump Start soft expulsions was significantly moderated by changes in center discipline policies, such that Jump Start was effective at reducing soft expulsions only when discipline policies showed meaningful improvement. Conclusions: Infant and early childhood consultation models, such as Jump Start, show promise in reducing exclusionary discipline practices, especially when implementation improves discipline policies.

Keywords: infant and early childhood mental health consultation; early care and education; exclusionary discipline practices; suspensions; expulsions; soft expulsions

1. Introduction

National data indicate that approximately 250 preschool children are suspended or expelled from classrooms each day in the United States [1]. These exclusionary discipline practices have been recognized as an ongoing problem in early care and education (ECE) settings and occur at rates exceeding those observed in K-12 education, with preschoolers expelled more than three times more than older students [2,3]. More specifically, Hispanic and Black preschool boys account for 66% of suspension and expulsion incidents, despite comprising only 46% of enrollment [3,4]. Exclusionary discipline in early childhood is associated with a range of adverse outcomes across life, including later school suspension and expulsion, academic failure, school dropout, and increased risk of justice system involvement, which collectively contribute to the "preschool to prison pipeline" and widen existing racial and socioeconomic inequities [2,5].

Beyond traditional suspensions and expulsions, ECE settings frequently rely on soft expulsion, an informal and often undocumented form of exclusionary discipline [6,7]. Soft expulsion refers to practices that implicitly push children out of the childcare center without formally mandating removal, particularly in contexts where expulsion is discouraged or prohibited [6]. Common examples include repeatedly sending children home early, removing children from their regular educational environment, reducing attendance hours, frequently contacting caregivers to pick up a child, or suggesting that a program is not a "good fit," ultimately leaving families with little choice but to withdraw their child [6]. Because these practices are rarely captured in administrative data, the prevalence of exclusionary discipline practices in early childhood is likely underestimated, obscuring the full scope of the problem [7]. Importantly, evidence suggests that soft expulsion persists even in programs with explicit non-expulsion policies, often reflecting limited staff capacity, insufficient behavioral supports, and beliefs that some children's behaviors cannot be effectively managed within the existing classroom structures [6].

Infant and early childhood mental health consultation (IECMHC) is an evidence-based approach that promotes children's social–emotional development and helps to reduce suspension and expulsion rates in ECE settings [8–10]. IECMHC utilizes a multifaceted approach (program-, classroom-, and child-level consultations) in which mental health consultants (MHC) build ECE directors' and classroom teachers' capacity to effectively address children's social and emotional needs through positive and supportive interactions in ECE settings [8,11]. Rather than providing direct therapeutic services to children, IECMHC focuses on strengthening adult skills, relationships, and systems that support young children's social–emotional and behavioral health [12,13]. At its core, IECMHC is designed to facilitate meaningful shifts in early childhood professionals' beliefs, attitudes, and practices, with increasing emphasis on equity-informed approaches that encourage reflection on

https://doi.org/10.3390/children13030328

bias and promote caregiving practices that are attentive to race, gender, language, and class [5,11].

Building on the Georgetown Framework for IECMHC [14], the Jump Start (JS) Early Childhood Consultation program is a multi-tiered IECMHC model operating at the program, classroom, and child levels. It is delivered by trained MHCs to strengthen ECE centers' capacity to support children with challenging behaviors and to enhance behavior management practices before issues escalate to traditional suspension/expulsion or soft expulsion. JS places explicit emphasis on center-wide discipline and expulsion policies, with the goal of reducing reliance on exclusionary discipline practices and promoting developmentally appropriate, equity-informed approaches to support behavior. The Georgetown IECMHC framework posits that consultation may influence exclusionary discipline practices both through classroom-level behavior management and through shifts in center-wide policies [14]. Changes in discipline and expulsion policies specifically may influence the strength of the association between JS program duration and reductions in exclusionary practices. While IECMHC models have demonstrated effectiveness in reducing exclusionary discipline practices [9,15,16], limited research has examined associations between changes in center-level policies and both traditional suspensions/expulsions and soft expulsions. Therefore, the purpose of the present study was to examine: (1) the relationships between center-level exclusionary discipline practices, JS participation, and related changes in discipline and expulsion policies; (2) changes in traditional suspensions/expulsions and soft expulsions from baseline to post-program assessment; and (3) whether changes in discipline and expulsion policies moderated the association between the JS participation duration and post-program exclusionary discipline practices.

2. Materials and Methods

2.1. Participants

This retrospective analysis included 270 ECE centers across Miami-Dade County that received an average of 7 months of JS services during the 2022–2023, 2023–2024, or 2024–2025 academic years. ECE programs were identified and recruited from a list provided by the funding agency, The Children's Trust. Total center enrollment ranged from 3 to 288 children, with a total of 16,021 children across all centers. On average, each center included 61 children, 5 classrooms, and 11 staff members. Children were racially (61% White, 21% Black, 12% multiracial) and ethnically diverse (69% Hispanic and 3% Haitian), and gender was evenly distributed (49% female). On average, center directors reported that 11% of children at their center had special needs. Over one third of children were enrolled in a school readiness or childcare subsidy program (36%), indicating lower socioeconomic status, and more than half of the ECE centers were enrolled in the county's Thrive by Five Initiative, a local program designed to improve access to high-quality ECE centers for children in high-poverty areas [17]. The University of Miami's Institutional Review Board approved this retrospective review of JS data to assess the program's effectiveness and inform future program improvements (IRB study number 20251177).

2.2. Measures

2.2.1. Demographics

During the enrollment phase, center directors completed both an intake form and a program demographics form, which collected information about their professional background, ECE center characteristics, and the children served.

https://doi.org/10.3390/children13030328

2.2.2. Frequency of Exclusionary Discipline

Incidences of exclusionary discipline practices were assessed via director self-report at baseline and immediately after completing JS using the Health Environment Rating Scale–Program (HERS-P) Self-Assessment. MHCs were trained to follow standardized instructions when administering the measure, to ensure consistent interpretation of items across participating directors. Respondents reported the number of traditional suspensions/expulsions that occurred at their center in the last 12 months, which was consistent with how formal exclusionary incidents are typically tracked and reported at the program level [2,3]. To assess the frequency of soft expulsions, directors were asked: "On average, how many times during the week are children that exhibit challenging behaviors removed from a classroom and sent to the office or another classroom?" This behavioral indicator is consistent with prior research operationalizing soft expulsion through reports of classroom removal practices [7], and weekly frequency was used because such practices are informal and not systematically documented in administrative records [6,7]. Because the JS duration averaged approximately 7 months, the differing time frames were intended to align with standard annual documentation of traditional suspensions/expulsions, while capturing more proximal changes in weekly classroom practices related to soft expulsions.

2.2.3. Discipline and Exclusion Policies

The Health Environment Rating Scale–Program (HERS-P) is a 30 min interview tool developed by the study team that examines the adoption of center-wide policies related to the JS model. Similar to the team's Health Environment Rating Scale–Classroom observation measure [18], the HERS-P aligns with core national standards, specifically those set by the National Association for the Education of Young Children [19] and Caring for Our Children [20], and covers four domains: safety, trauma-informed behavioral supports, self-care, and communication. Interviews were conducted with center directors by MHCs who completed a 5-h HERS-P training and achieved at least 80% agreement on ratings of recorded HERS videos to establish interrater reliability. For the purposes of this study, analyses focused on the total scores from two subdomains within behavioral supports: (1) expulsion and suspension policy (e.g., "The program has an expulsion and suspension policy that eliminates or severely limits expulsion, suspension, or other exclusionary disciplines") and (2) discipline policy (e.g., "The program has established developmentally appropriate social–emotional discipline practices and intervention procedures"). The HERS-P is scored on a 7-point Likert scale ranging from "far below standards" (1) to "exceeds standards" (7). Both subscales of the HERS-P exhibited acceptable internal reliability (Suspension and Expulsion Policy: $\alpha = 0.81$–0.87; Discipline Policy: $\alpha = 0.71$–0.82).

2.3. *Procedures*

The JS program is a multi-tiered IECMHC model implemented in ECE centers across Miami-Dade County. To determine readiness for collaborative consultation, centers were initially screened for recent participation in formal preschool expulsion-reduction initiatives. In addition, MHCs conducted semi-structured interviews with center directors prior to enrollment that covered program leadership, discipline practices, and capacity to engage in reflective consultation and programmatic change. Centers were considered ready for participation when directors indicated no prior participation in expulsion-reduction initiatives and demonstrated openness to partnership and willingness to engage in consultation activities, including completing required assessments and allowing MHC access to staff and classrooms for observation. Demographic data were collected from centers that met readiness criteria and enrolled in the program. Next, center directors and selected classroom teachers completed informed consent and demographic questionnaires. Two teachers

https://doi.org/10.3390/children13030328

per center were chosen by the center director to also collaborate with the MHC. Selections were made based on the teachers' needs for growth, classroom challenges, and their own motivation to participate.

JS services were offered for up to 10 months per center. MHCs were assigned caseloads of approximately four centers and spent one full day per week at each site. Consultation activities included program-level consultation, with administrators focused on policies, practices, and exclusionary discipline prevention strategies (approximately two hours per visit); classroom-level consultation focused on supporting teachers' capacity to promote children's social–emotional development and address challenging behaviors using developmentally appropriate strategies (approximately four hours per visit); and child-level consultation with caregivers and teachers of identified children with persistent behavioral challenges (that placed them at risk for traditional suspension/expulsion or soft expulsion) to help them reflect on the meaning of behaviors and develop individualized strategies to strengthen the protective factors (approximately two hours per visit). Two child-level cases per center were selected by directors and teachers with parents' consent.

Consultant caseload assignment was guided by cultural and language compatibility whenever possible. Centers reported the primary language(s) spoken by staff and the racial and ethnic composition of teachers and administrators at intake. Services were provided in English, Spanish, and Haitian Creole, and consultants were matched to centers based on shared language, cultural background, or prior professional relationships to facilitate engagement and implementation.

2.4. Analysis

Descriptive and correlational analyses explored relationships among exclusionary discipline practices at baseline, JS program duration, discipline and expulsion policy changes, and exclusionary discipline practices after completing JS. Two Wilcoxon signed-rank tests were conducted to evaluate pre- and post-JS differences in traditional suspensions/expulsions and soft expulsions. In addition, four moderation analyses were conducted using the PROCESS macro with 5000 bootstrap samples to examine whether the relationships between JS program duration and post-program exclusionary discipline practices (traditional suspensions/expulsions and soft expulsions) were moderated by changes in ECE center policies (discipline and expulsion policies). Total center enrollment and frequency of exclusionary discipline practices prior to JS were included as covariates in each model.

3. Results

3.1. Preliminary Analyses

At baseline, center directors reported an average of 0.25 traditional suspensions/expulsions in the last 12 months, with 16% of centers reporting at least one traditional suspension/expulsion in the last 12 months. In addition, center directors reported an average of 2.22 soft expulsions per week, with 45% of centers reporting at least one soft expulsion per week. On average, ECE centers implemented JS for 7 months between baseline and post-program assessment. During this time, center directors reported positive changes in their discipline (M = 1.06) and expulsion (M = 0.95) policies. Significant associations emerged between changes in discipline and expulsion policies ($r = 0.49$, $p < 0.001$). After completing JS, center directors reported an average of 0.13 traditional suspensions/expulsions in the last 12 months, with 11% of ECE centers reporting at least one traditional suspension/expulsion in the last 12 months. In addition, center directors reported an average of 0.58 soft expulsions per week, with 33% of ECE centers reporting at least one soft expulsion per week. Significant associations emerged between traditional

 https://doi.org/10.3390/children13030328

suspensions/expulsions and soft expulsions after completing JS ($r = 0.15$, $p = 0.02$). See Table 1.

Table 1. Descriptive statistics and associations between study variables.

	M	SD	1	2	3	4	5	6
Baseline								
1. Traditional Expulsions	0.25	0.78						
2. Soft Expulsions	2.22	13.76	−0.01					
Jump Start								
3. Duration in Months	7.10	2.32	0.01	−0.04				
4. Discipline Policy Change	1.06	0.88	0.05	−0.03	0.10			
5. Expulsion Policy Change	0.95	0.90	0.12	−0.06	0.13	0.49 **		
Post-Program								
6. Traditional Expulsions	0.13	0.38	0.25 **	0.00	−0.07	0.01	0.08	
7. Soft Expulsions	0.58	1.27	0.00	0.14 *	−0.12	−0.02	0.04	0.15 *

Note. Center enrollment at baseline included as covariate. ** $p < 0.01$. * $p < 0.05$.

3.2. Effect of Jump Start on Exclusionary Discipline Practices

A first Wilcoxon signed-rank test was conducted to evaluate the effect of JS on the frequency of traditional suspensions/expulsions. Results indicated a statistically significant difference in the frequency of traditional suspensions/expulsions before and after JS, $Z = -2.56$, $p = 0.01$. Traditional suspensions/expulsions decreased from 0.25 (Md < 1) to 0.13 (Md < 1) in the last 12 months, suggesting that JS significantly reduced traditional suspensions/expulsions.

A second Wilcoxon signed-rank test was conducted to evaluate the effect of JS on the frequency of soft expulsions. The results indicated a statistically significant difference in the frequency of soft expulsions before and after JS, $Z = -4.79$, $p < 0.001$. Soft expulsions decreased from 2.22 (Md < 1) to 0.58 (Md < 1) per week, suggesting that JS significantly reduced soft expulsions.

3.3. Moderation of Jump Start on Exclusionary Discipline Practices by Policy Changes

The overall model examining the moderation of JS program duration and soft expulsions after completing JS by changes in discipline policies, controlling for total center enrollment and soft expulsions at baseline, was statistically significant, $R^2 = 0.05$, $F(5, 240) = 2.63$, $p = 0.02$. A significant interaction emerged between the JS program duration and changes in discipline policies, $b = -0.08$, $SE = 0.04$, $t(240) = -1.98$, $p < 0.05$. This interaction accounted for a significant portion of the variance in soft expulsions, $\Delta R^2 = 0.02$, $F(1, 240) = 3.93$, $p < 0.05$. See Figure 1.

Simple slopes were analyzed to probe this interaction at three levels of discipline policy change (16th, 50th, and 84th percentiles). When there was no change to discipline policies (ΔPolicy = 0), the JS program duration did not significantly predict post-program soft expulsions ($b = 0.03$, $p = 0.58$). Similarly, at low levels of policy change (ΔPolicy = 1), the effect remained non-significant ($b = -0.05$, $p = 0.17$). However, when changes to discipline policies were high (ΔPolicy = 2), a longer JS program duration was significantly associated with lower post-program soft expulsions ($b = -0.13$, $SE = 0.05$, $t(240) = -2.47$, $p = 0.01$). These results suggest that JS is effective at reducing soft expulsions when implementation significantly improves center discipline policies. See Table 2 and Figure 2.

https://doi.org/10.3390/children13030328

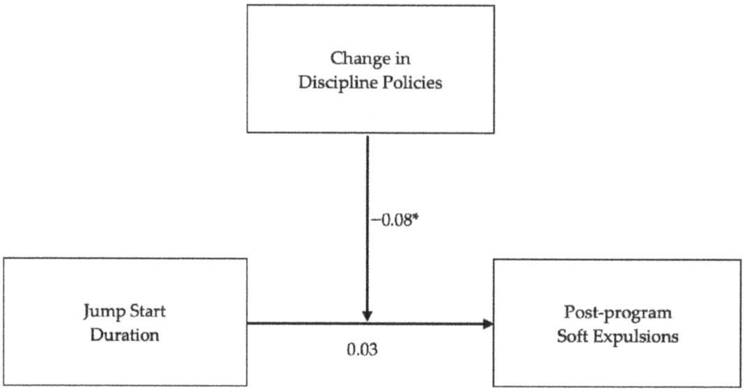

Figure 1. Moderation of duration on soft expulsions by discipline policy change. *Note.* Center enrollment and soft expulsions at baseline included as covariates. * $p < 0.05$.

Table 2. Conditional effects of Jump Start duration on post-program soft expulsions at different levels of discipline policy change.

Level of Policy Change	Effect (*b*)	*SE*	*t*	*p*	95% CI
Low (0)	0.03	0.05	0.55	0.58	[−0.08, 0.14]
Moderate (1)	−0.05	0.04	−1.37	0.17	[−0.12, 0.02]
High (2)	−0.13	0.05	−2.47	0.01	[−0.23, −0.03]

Note. CI = confidence interval. Center enrollment and soft expulsions at baseline included as covariates.

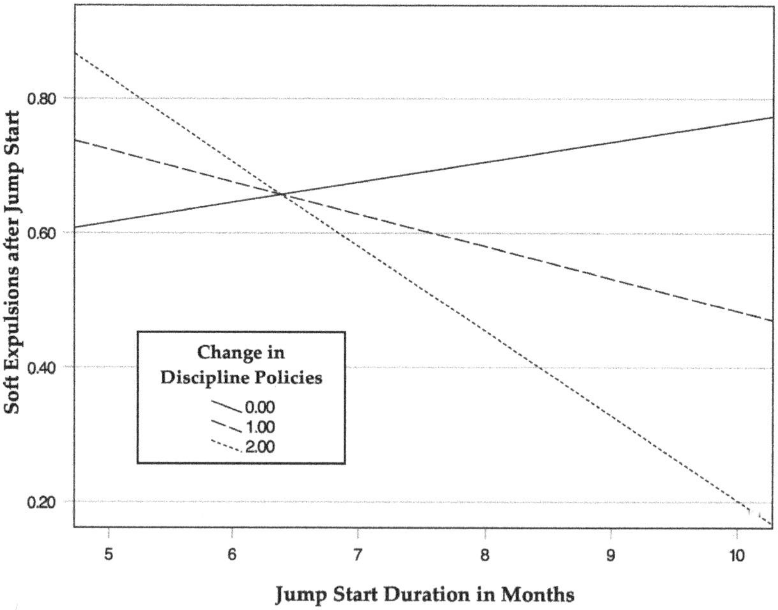

Figure 2. Conditional effects of Jump Start duration on soft expulsions at different levels of discipline policy change. Note. Center enrollment and soft expulsions at baseline included as covariates.

The three models examining JS program duration and soft expulsions by changes in expulsion policies, JS program duration and traditional suspensions/expulsions by changes

https://doi.org/10.3390/children13030328

in discipline policies, and JS program duration and traditional suspensions/expulsions by changes in expulsion policies did not include significant moderation effects (all $p > 0.12$).

4. Discussion

The current study examined whether participation in JS, a multi-tiered IECMHC model delivered at the program, classroom, and child levels, was associated with changes in exclusionary discipline practices within ECE centers. Although JS includes consultation activities across multiple levels, the current analyses focused specifically on center-level outcomes, including traditional suspensions/expulsions and soft expulsions, as well as changes in center-wide discipline and expulsion policies. In particular, this study explored how changes in ECE center policies influenced the relationship between the JS program duration and exclusionary discipline practices after completing JS. Overall, findings suggest that participation in JS was associated with meaningful reductions in exclusionary discipline practices, with the most robust and consistent associations observed for soft expulsion practices.

Consistent with prior research demonstrating that IECMHC models are effective in reducing suspension and expulsion in early childhood settings, participating centers reported relatively low rates of traditional suspensions/expulsions at baseline, with modest changes observed following JS participation [9,15,16]. Much of this prior work, including studies by Gilliam et al., has focused solely on traditional suspensions and expulsions, which are increasingly rare in jurisdictions with strong policy guidance discouraging these practices [2,5,15]. In the present study, soft expulsion practices were substantially more prevalent at baseline than traditional suspension/expulsion practices and demonstrated significant reductions at post-program assessment. Nearly half of participating ECE centers reported engaging in at least one soft expulsion per week at baseline, whereas only one-third reported any soft expulsions following completion of JS. These findings align with emerging evidence suggesting that soft expulsion practices are both more prevalent and potentially more harmful to children's continuity of care and social–emotional development than traditional exclusionary discipline practices, particularly because it operates outside formal reporting and accountability mechanisms. Together, these results suggest that soft expulsion may represent a more sensitive and prevalent indicator of exclusionary discipline practices in contexts where formal suspension and expulsion are discouraged or restricted.

The observed reductions in soft expulsion are especially notable, given the growing evidence that such practices often replace traditional exclusionary discipline practices despite explicit non-expulsion policies [7]. Soft expulsion practices are often undocumented and therefore remain largely underreported in administrative data, despite their potential to disrupt children's continuity of care and disproportionately affect children from historically marginalized backgrounds. The present findings underscore the importance of explicitly measuring soft expulsion when evaluating efforts to reduce exclusionary discipline practices in ECE settings and suggest that program-level consultation may play a critical role in addressing these less visible practices.

Moderation analyses further indicated that reductions in soft expulsion were strongest within ECE centers that demonstrated meaningful improvements in discipline policies during participation in JS. Specifically, longer engagement in JS was associated with fewer soft expulsions only when centers exhibited improved discipline policies. In contrast, no such associations were observed among centers with minimal or no policy change. This pattern suggests that multi-tiered consultation efforts may be most effective at reducing informal exclusionary discipline practices when classroom- and child-level supports are reinforced by clear, developmentally appropriate, and equity-informed center-level policies. No significant moderation effects were observed for traditional suspensions/expulsions.

https://doi.org/10.3390/children13030328

This likely reflects the low rates of traditional suspension/expulsion observed at baseline, which may have limited variability and statistical power to detect moderation effects. Together, these findings suggest that while IECMHC models may contribute to maintaining low levels of traditional exclusionary discipline practices, their greatest impact may be in reducing soft expulsion practices that remain more common and less regulated.

4.1. Limitations

Several limitations should be considered when interpreting these findings. First, exclusionary discipline practices were self-reported by center directors and may be subject to recall bias or social desirability bias. Although a standardized self-assessment measure was used, some variability in how exclusionary discipline practices, particularly soft expulsion, were interpreted or categorized may remain, with potential implications for construct validity. In addition, HERS-P interviews were conducted by the same MHCs providing JS services, introducing the potential for expectancy bias.

Second, the baseline rates of traditional suspensions/expulsions were relatively low. This likely reflects the broader policy and practice context of Miami-Dade County, where coordinated, county-wide initiatives aimed at reducing preschool suspensions and expulsions have been in place for several years. Efforts led by the Early Learning Coalition and partnerships among local early childhood and mental health organizations may have contributed to the widespread adoption of non-expulsion policies prior to centers' participation in JS [21]. These same broader system-level efforts should be considered when interpreting changes during the study period.

Another important consideration is the study's retrospective design and lack of a control group, which limits the ability to draw causal inferences about program effects. To reduce potential confounding from prior intervention exposure, centers were screened prior to enrollment for recent participation in formal preschool expulsion-reduction initiatives, and those with intensive prior involvement were not eligible. This step was intended to strengthen the internal validity and minimize contamination from previous targeted supports. As such, observed improvements should be interpreted as associations occurring during JS implementation, rather than definitive evidence of the program effects. Lastly, overall participation in JS may have influenced center policies, limiting conclusions regarding the temporal ordering of program duration, policy changes, and reductions in soft expulsions.

4.2. Future Directions

Despite these limitations, the present findings have important implications for policy and practice. The results suggest that IECMHC models that explicitly target program-level discipline policies may be particularly effective in reducing soft expulsions, an exclusionary discipline practice that is often underreported and difficult to track. Interventions that focus solely on reducing traditional suspensions and expulsions may underestimate the true scope of exclusionary discipline practices and overlook opportunities for meaningful change.

Future research should continue to refine the measurement of soft expulsion and explore how policy implementation fidelity, rather than policy presence alone, shapes exclusionary discipline practices. In addition, further research is needed to examine how center-level policy changes interact with classroom- and child-level consultation processes to influence exclusionary discipline practices over time and whether reductions in soft expulsions narrow disparities across centers serving historically marginalized populations. Lastly, researchers should explore policy changes as potential mediators between pro-

 https://doi.org/10.3390/children13030328

gram participation and exclusionary discipline practices to further clarify the conceptual pathways underlying soft expulsions.

5. Conclusions

This study found that participation in JS may have been associated with reductions in exclusionary discipline practices at the center level, with the largest changes observed in soft expulsions. Although traditional suspensions and expulsions were relatively uncommon, soft expulsions were prevalent at baseline and occurred less frequently following program participation. Reductions in soft expulsion were observed primarily among centers that demonstrated meaningful improvements in discipline policies, suggesting that policy change may play an important role in shaping informal exclusionary practices. Together, these findings highlight the value of examining soft expulsion as a key indicator of exclusionary discipline practices and suggest that IECMHC models that attend to center-wide discipline policies may be well positioned to support more inclusive early care and education environments.

Author Contributions: Conceptualization, R.N., Y.A. and M.D.D.; methodology, Y.A. and M.D.D.; formal analysis, M.D.D.; investigation, R.N. and M.S.; resources, R.S.; data curation, Y.A. and M.D.D.; writing—original draft preparation, Y.A., M.D.D., A.R., M.S. and R.N.; writing—review and editing, B.C., E.E., T.R. and R.S.; visualization, M.D.D.; supervision, A.R., M.S., B.C., E.E. and T.R.; project administration, A.R., M.S. and R.N.; funding acquisition, R.N. All authors have read and agreed to the published version of the manuscript.

Funding: This research was funded by The Children's Trust of Miami-Dade County, Florida, grant number GR027724.

Institutional Review Board Statement: The study was conducted in accordance with the Declaration of Helsinki and approved by the Institutional Review Board of the University of Miami (protocol code 20251177 and approved on 30 October 2025).

Informed Consent Statement: Informed consent was not required due to the retrospective nature of the study and the use of de-identified program data for secondary analysis.

Data Availability Statement: Requests for data can be sent to the corresponding author.

Acknowledgments: We would like to thank the early care and education centers in Miami-Dade County whose participation in the Jump Start program made this work possible.

Conflicts of Interest: Bianca Caceres was employed by the company Community Health of South Florida. Elizabeth Erban was employed by the company Family Central. The remaining authors declare that the research was conducted in the absence of any commercial or financial relationships that could be construed as a potential conflict of interest.

Abbreviations

The following abbreviations are used in this manuscript:

ECE	Early Care and Education
HERS-P	Health Environment Rating Scale-Program
IECMHC	Infant and Early Childhood Mental Health Consultation
IRB	Institutional Review Board
JS	Jump Start
MHC	Mental Health Consultant

References

1. New Data Reveal 250 Preschoolers Are Suspended or Expelled Every Day. Available online: https://www.americanprogress.org/article/new-data-reveal-250-preschoolers-suspended-expelled-every-day/ (accessed on 1 December 2025).

https://doi.org/10.3390/children13030328

2. Gilliam, W.S. *Prekindergartners Left Behind: Expulsion Rates in State Prekindergarten Systems*; Foundation for Child Development: New York, NY, USA, 2005.
3. Williams, P.G.; Yogman, M. Addressing early education and child care expulsion. *Pediatrics* **2023**, *152*, e2023064049. [CrossRef] [PubMed]
4. Zeng, S.; Corr, C.P.; O'Grady, C.; Guan, Y. Adverse childhood experiences and preschool suspension expulsion: A population study. *Child Abus. Negl.* **2019**, *97*, 104149. [CrossRef] [PubMed]
5. Meek, S.E.; Gilliam, W.S. *Expulsion and Suspension in Early Education as Matters of Social Justice and Health Equity*; NAM Perspectives; National Academy of Medicine: Washington, DC, USA, 2016. [CrossRef]
6. Murphy, K.; Giordano, K.; Hoffstein-Rahmey, D.; Reizin, R.; Coyne, A. Soft expulsion: A solution when there are no other options? *Child. Youth Serv. Rev.* **2024**, *158*, 107441. [CrossRef]
7. Steyer, L.; Gleit, R.; Hawkins, C.; Provençal, M.; Pearman, F.; Obradović, J. Informal exclusionary discipline practices in US schools: Recent evidence and policy considerations. *Soc. Policy Rep.* **2025**, *38*, 3–22. [CrossRef]
8. Perry, D.F.; Allen, M.D.; Brennan, E.M.; Bradley, J.R. The evidence base for mental health consultation in early childhood settings: A research synthesis addressing children's behavioral outcomes. *Early Educ. Dev.* **2010**, *21*, 795–824. [CrossRef]
9. Silver, H.C.; Davis Schoch, A.E.; Loomis, A.M.; Park, C.E.; Zinsser, K.M. Updating the evidence: A systematic review of a decade of Infant and Early Childhood Mental Health Consultation (IECMHC) research. *Infant Ment. Health J.* **2023**, *44*, 5–26. [CrossRef] [PubMed]
10. Trivedi, P.A.; deMonsabert, J.; Horen, N.M. *Infant and Early Childhood Mental Health Consultation: Overview of Research, Best Practices, and Examples*; SRI International: Menlo Park, CA, USA, 2021.
11. Duran, F.; Hepburn, K.; Irvine, M.; Kaufmann, R.E.; Anthony, B.; Horen, N.M.; Perry, D.F. *What Works? A Study of Effective Early Childhood Mental Health Consultation Programs*; Georgetown University Center for Child and Human Development: Washington, DC, USA, 2009.
12. Center of Excellence for Infant and Early Childhood Mental Health Consultation. Status of the Evidence for Infant and Early Childhood Mental Health Consultation. Available online: https://www.iecmhc.org/documents/CoE-Evidence-Synthesis.pdf (accessed on 1 December 2025).
13. Natale, R.; Kolomeyer, E.; Futterer, J.; Mahmoud, F.D.; Schenker, M.; Robleto, A.; Horen, N.; Spector, R. Infant and early childhood mental health consultation in a diverse metropolitan area. *Infant Ment. Health J.* **2022**, *43*, 440–454. [CrossRef] [PubMed]
14. Hunter, A.; Davis, A.; Perry, D.F.; Jones, W. *The Georgetown Model of Early Childhood Mental Health Consultation: For School-Based Settings*; Georgetown University Center for Child and Human Development: Washington, DC, USA, 2016.
15. Gilliam, W.S.; Maupin, A.N.; Reyes, C.R. Early childhood mental health consultation: Results of a statewide random-controlled evaluation. *J. Am. Acad. Child Adolesc. Psychiatry* **2016**, *55*, 754–761. [CrossRef] [PubMed]
16. Newland, R.; Silver, R.B.; Herman, R.; Hartz, K.; Coyne, A.; Seifer, R. Child-focused infant and early childhood mental health consultation: Shifting adult attributions to reduce the risk for preschool expulsion. *Infant Ment. Health J.* **2024**, *45*, 249–262. [CrossRef] [PubMed]
17. Thrive by Five. Available online: https://www.thechildrenstrust.org/about-us/what-we-fund/thrive-by-5/ (accessed on 26 January 2026).
18. Futterer, J.; Mullins, C.; Bulotsky-Shearer, R.J.; Guzmán, E.; Hidalgo, T.; Kolomeyer, E.; Howe, E.; Horen, N.; Sanders, L.M.; Natale, R. Initial validation of the Health Environment Rating Scale–Early Childhood Consultation–Classroom (HERS-ECC-C). *Infant Ment. Health J.* **2024**, *45*, 449–463. [CrossRef] [PubMed]
19. National Association for the Education of Young Children. Professional Standards and Competencies for Early Childhood Educators. Available online: https://www.naeyc.org/sites/default/files/globally-shared/downloads/PDFs/resources/position-statements/naeyc_standards_and_competencies_ps_final.pdf (accessed on 1 December 2025).
20. American Academy of Pediatrics; American Public Health Association; National Resource Center for Health and Safety in Child Care and Early Education. *Caring for Our Children, National Health and Safety Performance Standards*, 4th ed.; Guidelines for Early Care and Education Programs; American Academy of Pediatrics: Itasca, IL, USA, 2019. [CrossRef]
21. Florida Department of Education; Office of Early Learning. *Position Statement on Expulsion and Suspension Prevention in Early Childhood Settings*; Florida Department of Education: Tallahassee, FL, USA, 2014.

https://doi.org/10.3390/children13030328

children

MDPI

Article

The Relations Between Sensory Modulation, Hyper Arousability and Psychopathology in Adolescents with Anxiety Disorders

Ginan Hammud [1], Ayelet Avital-Magen [2], Hiba Jabareen [2], Reut Adler-Tsafir [1] and Batya Engel-Yeger [1,*]

[1] Department of Occupational Therapy, Faculty of Social Welfare and Health Sciences, University of Haifa, Haifa 3498838, Israel; ghammud@campus.haifa.ac.il (G.H.); reutadler@gmail.com (R.A.-T.)
[2] Child and Adolescent Mental Health Clinic, Haemeq Medical Center, Afula 1834111, Israel; ayelet_av@clalit.org.il (A.A.-M.); hibajaba@clalit.org.il (H.J.)
* Correspondence: bengel@univ.haifa.ac.il

Abstract: Background: Sensory modulation may play a significant role in psychiatric conditions, including anxiety, and explain arousability levels, behavioral disorders, and functional deficits. Yet, studies about sensory modulation in adolescents with anxiety disorders are scarce. Purpose: To profile the prevalence of sensory modulation difficulties (SMDs) in adolescents with anxiety and examine their relations to arousability and psychopathology. The study compared adolescents with anxiety disorders to healthy controls using physiological measures and self-reports that reflect daily life scenarios. Then, the study examined the relationship between SMDs, arousability, and psychopathological severity in the study group. Method: Participants were 106 adolescents, aged 10.5–18 years and their parents. The study group included 44 participants diagnosed with anxiety disorder by psychiatrists. The control group included 62 healthy participants matched by age and gender to the study group. Parents completed the demographic questionnaire and the Child Behavior Checklist (CBCL). The adolescents completed The Revised Children's Manifest Anxiety Scale (RCMAS) and the Adolescent/Adult Sensory Profile (AASP) and underwent the electrodermal activity (EDA) and pulse rate tests while listening to extreme sensory stimuli of auditory startles. Results: Based on AASP, the study group had a higher prevalence of SMDs expressed in lower sensory seeking, difficulties in registering sensory stimuli, and higher sensory sensitivity and avoidance. The study group presented higher arousability while listening to the startles as manifested in higher heart rate and EDA responses. The physiological results correlated with SMD levels measured by the AASP self-reports. SMDs correlated with psychopathological severity. Conclusions: SMDs may characterize adolescents with anxiety disorders and impact their arousability, symptoms severity, and daily functioning. Therefore, sensory modulation should be evaluated using both self-reports (to reflect implications in real life from patients' own voices) along with objective measures to explain daily behaviors by underlying physiological mechanisms. This may focus intervention towards better health, function, and development.

Keywords: anxiety disorders; sensory modulation; hyper-arousability; daily life

1. Introduction

Anxiety disorders are prevalent among 20.5% of children and adolescents around the world [1]. As outlined in The Diagnostic and Statistical Manual of Mental Disorders (DSM-V), anxiety disorders include social anxiety disorder (SAD), specific phobias, generalized anxiety disorder (GAD), panic disorder (PD), separation anxiety, selective mutism,

https://doi.org/10.3390/children12020187

and agoraphobia. Among children and adolescents under the age of 18, the definition of anxiety disorder is relevant if symptoms last for at least six months [2]. Anxiety disorders often coexist with other psychiatric conditions, like attention deficit hyperactivity disorder (ADHD), major depression, somatic symptom disorders, personality disorders, and substance abuse disorders [3,4]. The symptoms of anxiety disorders include worry about the future, obsessive and repetitive thoughts, and excessive fear [2,5] that physiologically are manifested in hyper-arousability and elevated sympathetic activity (increased heart rate, shortness of breath, sweating, tremor, chest pain, nausea, etc.). Hyper-arousability may be triggered by a disproportionate response to stimuli from the environment that are perceived as threatening [6]. This may explain the high prevalence of sensory modulation difficulties (SMDs), mainly sensory hypersensitivity in patients with anxiety disorders [7–10].

Sensory modulation refers to the ability of the central nervous system to modulate and organize sensory input in a graded manner according to environmental demands [11]. Difficulties in sensory modulation result in hypersensitivity or hyposensitivity and lead to unmodulated behavior and malfunction. One main model that refers to the implications of sensory modulation on behavior is Dunn's model. According to this model, the interaction between the neural threshold for sensory input and the strategies used for coping with the neural threshold (active/passive) defines behavioral and functional outcomes [11]. This interaction results in four profiles: sensory sensitivity (low neural threshold and passive strategy)—people who do not actively eliminate unpleasant stimuli; sensation avoidance (low neural threshold and an active strategy)—people who actively avoid unpleasant stimuli, such as avoiding walking on sand; low registration (high neural threshold and passive strategy)—people who need extensive intensity of sensory stimuli to recognize it, but do not actively seek rich sensory input (as sports activity), therefore, may miss information, and be less attentive; sensory seeking (high neural threshold and an active strategy)—people who actively seek for intense stimuli in their environment, as going to parties; coloring their rooms; do extreme sport [12]. One of the gold standard measures for sensory modulation is the Sensory Profile [13]. Based on the sensory profile, SMDs are prevalent among 57% of children with psychiatric disorders [14]. The significant benefit of the sensory profile is that it reflects SMDs as expressed in behavioral responses to sensory input in daily life scenarios.

Indeed, when the ability to modulate sensory information is impaired, it affects the individual's ability to properly perceive the environment and function in it. Therefore, anxiety and hyper-arousability may be elevated, as manifested in objective physiological measures such as electrodermal activity (EDA) [15], higher cortisol levels [16], and elevated heart rate [17–20]. In children, anxiety symptoms were manifested in physiological measures such as over-reactive skin conductance (EDA) [21]. Hence, the objective and subjective measures may explain unmodulated behaviors such as "fight or flight," impulsivity, distraction, nonadaptive behavior and daily dysfunction [22].

Hence, the complexity of the symptoms of anxiety disorders, the prevalent comorbidities of SMDs as well as their severe consequences on a child's/adolescent's daily life, serve the rationale for studying the interrelation between these factors in depth in order to improve intervention. This knowledge is especially critical in adolescence where development is essential for determining later human life and wellbeing. Therefore, this study aimed to (1) profile sensory modulation of adolescents with anxiety disorders, as compared to healthy controls, using objective and subjective measures and (2) among the study group, examine the relationship between sensory modulation and anxiety psychopathology.

Study hypotheses: (1) adolescents with anxiety disorders would have a significantly higher prevalence of SMDs as expressed in self-reports and in objective measures of sympathetic hyper-arousability than healthy controls; (2) among the study group—(a) SMDs according to self-reports would correlate with the objective manifestations and (b) SMDs would significantly correlate with severe psychopathology.

2. Materials and Methods

2.1. Participants

This study is a cross-sectional and correlative study. Based on the power calculation of MANOVA global effects, with f2 = 0.07 at a significance level of $p = 0.05$ and power = 0.80, this study was supposed to include 142 adolescents. However, due to various constraints, this study included 106 adolescents. The participants' age range was 10.5–18 years. The study group included 44 adolescents, 21 boys and 23 girls, with a mean age of 13.69 (SD = 1.99), diagnosed with anxiety disorder by psychiatrists, according to the DSM-5 and a clinical interview. All participants of the study group were patients of the Child and Adolescent Mental Health Clinic (blinded). The control group included 62 healthy adolescents, 33 boys and 29 girls, mean age = 13.39 (SD = 1.99), with normal development and emotional health, based on their parents' report and their Child Behavior Checklist (CBCL) normal scores (Achenbach and Rescorla, 2001) [23]. The study groups were matched by age ($t = -0.76$, $p = 0.80$) and gender (χ^2 (1) = 0.31, $p = 0.57$) but significantly differed in mother's education (χ^2 (1) = 15.66, $p < 0.001$) and in socioeconomic status (χ^2 (2) = 21.14, $p < 0.001$) (see Tables 1 and 2). Mothers' education and socioeconomic status parameters were significantly positively correlated. Therefore, statistical adjustments were performed and the mother's education was held as a covariant. Exclusion criteria: adolescents with unstable severe health/chronic illnesses who are taking pharmacological treatment that has a detrimental effect on the nervous system or addictive substances (beyond the psychiatric medications in the study group) and those with uncorrected sensory impairment (e.g., by glasses).

Table 1. Prevalence of anxiety disorders among the study group.

Psychiatric Diagnosis	Number of Participants (*n* = 44)	Percentile %
Anxiety	8	18.2
Anxiety and ADHD	12	27.3
Anxiety and ODD	2	4.5
Anxiety and depression	5	11.4
Anxiety and adaptive disorder	3	6.8
Anxiety, ADHD and ODD	3	6.8
Anxiety, ADHD, and adaptive disorder	2	4.5
Anxiety, ADHD, and depression	1	2.3
Anxiety, ADHD, and OCD	3	6.8
Anxiety, ADHD, and PTSD	2	4.5
Anxiety, adaptive and borderline personality disorders	1	2.3
Anxiety, depression, and personality disorder	1	2.3
Anxiety, ADHD, and tics	1	2.3

Note: ADHD = Attention deficit and hyperactivity disorder; ODD = Oppositional defiant disorder; OCD = Obsessive-compulsive disorder; PTSD = Post-traumatic stress disorder.

Table 2. Sociodemographic characteristics of both groups.

		Control Group (*n* = 62)		95% CIs		Study Group (*n* = 44)		95% CIs	
		n	Percentile %	Lower Bound	Upper Bound	*n*	Percentile %	Lower Bound	Upper Bound
Mother education	Elementary school	0	0	0	0.058	0	0	0	0.080
	High school	8	12.9	0.066	0.234	10	22.7	0.128	0.37
	Professional	1	1.6	0.002	0.085	11	25.0	0.145	0.394
	Academic	53	85.5	0.746	0.921	22	50.0	0.358	0.641

Table 2. *Cont.*

		Control Group (*n* = 62)		95% CIs		Study Group (*n* = 44)		95% CIs	
Father education	Elementary school	0	0	0	0.058	1	2.3	0.004	0.118
	High school	3	4.8	0.016	0.132	11	25	0.145	0.394
	Professional	6	9.7	0.045	0.195	10	22.7	0.128	0.37
	Academic	53	85.5	0.746	0.921	20	45.5	0.317	0.599
Socioeconomic status	Below average	2	3.2	0.008	0.110	13	29.5	0.181	0.442
	Average	6	9.7	0.045	0.195	10	22.7	0.128	0.37
	Above average	54	87.1	0.765	0.933	21	47.7	0.337	0.620

2.2. Measures

2.2.1. Subjective Measures—Self-Reports

Completed by Parents

A sociodemographic questionnaire.

The Child Behavior Checklist (CBCL) [23] was used to profile psychopathological symptoms. This standard measure screens behavioral and emotional problems in children and adolescents ages 4–18. This measure includes two sections: 20 items that measure social functioning and 113 items that measure emotional–behavioral problems (behavioral profile). For this study, the behavioral part was used. Answers are marked on a 0–2 scale (0 = not true for my child, 1 = somewhat/sometimes true, and 2 = very true or often true). Parents rate the items of problems based on the past six months. Raw scores are converted into T scores that indicate whether the participant exhibits abnormal behaviors with respect to norms by age and gender. A clinical cross-sectional score above 63, based on the total T-score, indicates clinical functioning, while scores between 60–63 suggest borderline functioning and scores below 60 are considered within the normal range. This measure demonstrates high test-retest reliability of 0.9, strong internal consistency of 0.72–0.97, and good criterion validity [23].

Completed by the Adolescents

The Revised Children's Manifest Anxiety Scale (RCMAS) [24] aims to profile anxiety levels in children and adolescents aged 6–19 years. The RCMAS includes 37 items, divided into three scales: The Physiological Scale consists of 10 items related to physical symptoms of anxiety, such as sleep problems, stomach issues and breathing difficulties. The Worry and Oversensitivity Scale consists of 11 items related to excessive, unfocused worrying and concern about external stressors. The Social Concern and Concentration Scale includes 7 items related to fears of social isolation and attention difficulties. The remaining 9 items comprise The Lie Scale, which measures social desirability and "fake good". Each item describes an experience, respondents note "Yes" or "No". If the response is "Yes" it indicates the presence of an anxiety symptom and it receives a score of 1. If the response is "No" it indicates the absence of an anxiety symptom and it receives a score of 0. The total score is calculated based on the three scales of the RCMAS which consist of 28 of the 37 items (the 9 items of the lie scale are excluded). Higher scores indicate higher levels of anxiety.

The RCMAS has good psychometric properties [24,25]. According to the standardization sample data [24], a T-score of 60 is equivalent to the cut-off score of 19 (M = 11.70, SD = 6.21).

Adolescent/Adult Sensory Profile (AASP) Questionnaire [26]: This 60-item self-report measures sensory modulation among individuals aged 11 and above. Respondents rate the frequency of their behavioral reactions to sensory experiences in everyday life (for example: "I'm uncomfortable wearing certain fabrics"; "I avoid elevators and/or escalators because

I dislike the movement"). Answers are scored on a 1–5 Likert scale (1 = almost never, 5 = almost always). The higher the score, the more extreme the sensory modulation pattern. The AASP has four outcome measures in accordance with the four quadrants of Dunn's model (see above): low registration, sensory seeking, sensory sensitivity, and sensation avoidance (15 items in each category). Five categories are identified in each sensory modulation pattern: "much less than most people," "less than most people," "similar to most people," "more than most people," and "much more than most people" [26]. In the present study, the five categories were combined into three: (1) "less than most people"—that is, one to two standard deviations below the norm, (2) "similar to most people"—within the norm, and (3) "more than most people"—one to two standard deviations above the norm. Sensory modulation norms exist for different age groups (11–17, 18–64, 65 and over). SMDs are defined for participants found in the "less" or "more than most people" ranges. Internal reliability was found to range from 0.64–0.78 [26]. Discriminant validity was found between groups in the psychiatric population [27].

2.2.2. Objective Measures

In order to reflect the physiological manifestations of SMDs, and based on the literature on SMDs, the following measures examine arousability/sympathetic reactivity to sensory stimuli of auditory startles:

An electrodermal activity test (EDA) [28] measures the skin conductivity (in units of micro-siemens) due to the activity of the sweat glands in the skin in response to a stressogenic stimulus. Two 5 mm electrodes (Mindlife, Jerusalem, Israel) are connected to the second and fourth fingertips of the non-dominant hand using a Velcro strap. The electrodes are connected to a sensor, receiver, and amplifier, providing 10 samples per second [28]. The experiment is divided into three phases, while the subject is sitting in front of a computer screen watching kaleidoscope shapes in order to neutralize stimuli distractions from the environment and to increase compliance which is often used in psychophysiology research [29]. The first phase lasts a minute (for acclimatization). In this phase, the subjects hear white noise through the headphones. The second phase—the startles session—includes the auditory startles which appear randomly, 40 times, for 3 min, at a volume of 90 dB with each stimulus lasting 30 ms. In the third and final phase, the relaxation session, white noise is heard again for one minute (allowing adjustment). The experiment lasts 5 min in total.

The outcome measures of the EDA test are (1) reaction time (in seconds) and amplitude height in response to the first startle; (2) time (in seconds) and height of the maximum amplitude during the 3 min of auditory startles—to measure maximal arousability; (3) the following measures reflected adaptation: time and height of minimal amplitude during the 3 min of auditory startles session; and (4) time and height of maximum and minimum amplitude at the time of relaxation session (third session).

A pulse oximeter (Contec Systems CMS50N; Qinhuangdao, China) is an additional measure of sympathetic activity reflected in heart rate in response to auditory startles. The pulse oximeter was attached to the other finger of the hand that is not connected to the EDA test during all EDA sessions.

2.3. Procedure

After obtaining ethics approval from the [blinded] and from the Helsinki Committee [blinded], adolescents and their parents from the study group were recruited via advertisements published at the Child and Adolescent Mental Health Clinic [blinded]. Participants in the control group were recruited through similar advertisements published in neighborhoods in the same geographic area.

Parents interested in participating with their children in this study were contacted by the study conductor and a meeting was scheduled with them at the clinic (study group) or at their homes (control group). In the meeting, the participants and their parents received a detailed explanation about the study and observing ethical rules declaration. The parents signed the informed consent form and completed the demographic questionnaire and the CBCL. Adolescents underwent the EDA test and the heart rate measurement. Then, they completed the RCMAS and the AASP with the study conductor (See Figure 1).

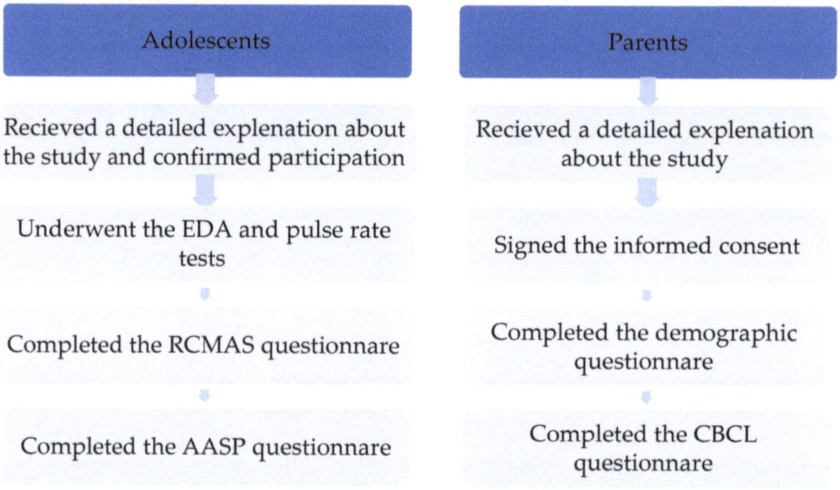

Figure 1. Experimental schedule completed by the participants and their parents.

All questionnaires were anonymous. Each questionnaire received a serial number without identifying details of the participants and the data was stored in a protected folder on the researcher's computer and all paper data was kept in a locked closet.

2.4. Statistical Analysis

Data was analyzed using the Statistical Package for the Social Sciences (SPSS)-28. After analyzing descriptive statistics and testing normality, the data distribution was found to be normal in all study measures, except for the CBCL scores in the control group. Therefore, chi-square analysis examined the prevalence of SMDs according to the AASP norms, clinical emotional behavior problems according to the CBCL and anxiety problems according to the RCMAS cutoff scores. T-tests examined the differences between the groups in socio-demographic parameters. For examining the first hypothesis, a univariate analysis of covariance (ANCOVA) examined the differences between the study groups in total scores of the RCMAS and heart rate. A multivariate analysis of covariance (MANCOVA) examined the differences between groups in the subscales scores of the RCMAS, AASP, and EDA scores. Mother's education was held as a covariate throughout the analyses. A Mann–Whitney test examined the differences between groups in CBCL scores. For examining the second hypothesis, the Pearson correlation test examined the correlations between the subjective (AASP) and objective (EDA and heartrate) measures and between sensory modulation and psychopathological symptoms severity according to the RCMAS and the CBCL. The significance level was $p \leq 0.05$.

In our study, the control group is larger in terms of the number of participants than the study group due to difficulties in recruiting participants for the study group. Furthermore, some participants were missing part of the data (for example, one of the questionnaires), so the degrees of freedom varied accordingly. In the context of the EDA test, for example, one participant in the study group had no responses to the test, and six participants showed no response to the first stimulus.

3. Results

Before examining the study's hypotheses, both groups were compared in their CBCL and RCMAS scores to verify inclusion criteria and provide the profile of each group.

3.1. The Differences in Emotional and Behavioral Modulation Psychopathology Between Both Groups According to the CBCL

According to the Mann–Whitney test, emotional and behavioral modulation problems were significantly more prevalent among the study group compared to the control group, as reflected in all CBCL scales and in the total score: the study group had greater difficulties in internalizing and externalizing symptoms. The study group had higher difficulties in anxiety symptoms, depression symptoms, and somatic complaints. Moreover, the study group showed significant behavioral problems such as social, thought and attention problems, rule-breaking, and aggressive behavior. (See Table 3, Figure 2)

Table 3. Comparing CBCL scores between groups according to the Mann–Whitney test.

	Study Group (n = 43) Mean + SD	Control Group (n = 60) Mean + SD	Mann-Whitney U	p
Anxious/Depressed	10.09 ± 4.7	2.12 ± 2.27	129	<0.001 ***
Withdrawn/Depressed	5.86 ± 3.79	0.92 ± 1.44	264	<0.001 ***
Somatic complaints	5.21 ± 4.54	0.90 ± 1.27	418	<0.001 ***
Social problems	6.14 ± 4.24	0.93 ± 1.56	204.5	<0.001 ***
Thought problems	6.37 ± 4.23	0.75 ± 1.09	217	<0.001 ***
Attention problems	8.88 ± 4.31	2.00 ± 2.53	224.5	<0.001 ***
Rule-breaking	3.65 ± 3.25	0.68 ± 0.98	516.5	<0.001 ***
Aggressive behavior	11.14 ± 6.77	2.40 ± 3.147	250.5	<0.001 ***
Other conditions	5.44 ± 3.41	1.67 ± 1.76	417.5	<0.001 ***
Internalization	68.33 ± 8.37	47.38 ± 7.95	77.5	<0.001 ***
Externalization	60.05 ± 8.66	43.53 ± 8.05	236.5	<0.001 ***
Total problems	65.79 ± 7.77	42.75 ± 10.18	86.5	<0.001 ***

Note: SD = standard deviation; *** $p \leq 0.001$.

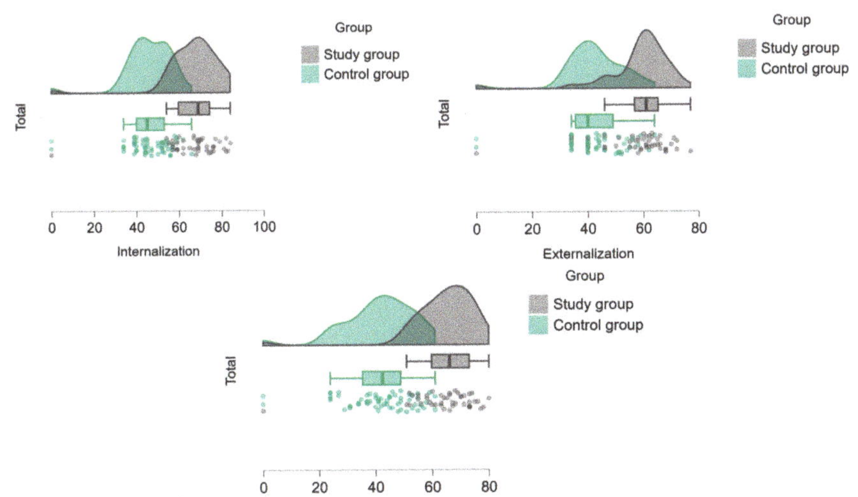

Figure 2. Differences between groups in CBCL scores.

According to chi-square analysis and the cutoff total CBCL score of 63, the study group had a significantly higher prevalence of clinical emotional and behavioral modulation problems (62.8%, CI95 48–76%) than the controls (0%, CI95 0–6%) ($\chi^2 = 51.059$, $p < 0.001$). The Fischer exact test indicated a significant difference between groups ($p < 0.001$).

3.2. The Differences in Anxiety Rates Between Both Groups According to the RCMAS

According to the MANCOVA and ANCOVA tests, anxiety symptoms were more prevalent among the study group, as reflected in all RCMAS scales: physiological, worry and over-responsivity, social concern, and concentration symptoms, except for the lie scale, and in total score, accordingly. The effect size for the observed anxiety symptoms variables was the partial eta square η_p^2 which ranged from 0.289 to 0.373, indicating large effect sizes (see Table 4, Figure 3).

Table 4. Comparing RCMAS scores between groups according to MANCOVA and ANCOVA tests.

	Study Group (n = 44) Mean + SE	Control Group (n = 62) Mean + SE	$F_{(1,106)}$	η_p^2	p	95% CIs Lower Bound	Upper Bound
Physiological	6.388 ± 0.355	3.289 ± 0.295	41.928	0.289	<0.001 ***	2.149	4.047
Worry and over-responsivity	8.239 ± 0.388	4.863 ± 0.323	41.479	0.287	<0.001 ***	2.336	4.416
Social concern and concentration	4.395 ± 0.269	1.865 ± 0.224	48.466	0.320	<0.001 ***	1.809	3.250
Lie	2.876 ± 0.361	2.781 ± 0.300	0.038	0.000	0.846	−0.871	1.061
Total anxiety score	18.298 ± 0.812	9.724 ± 0.675	61.244	0.373	<0.001 ***		

Note: SE = standard error; η_p^2 (partial η^2) = effect size; *** $p < 0.001$.

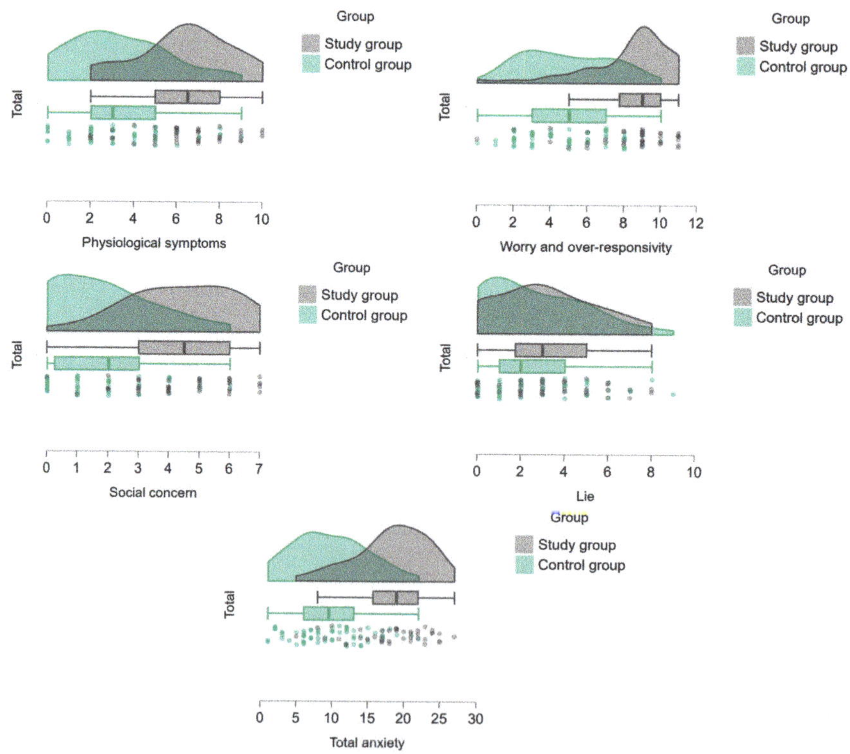

Figure 3. Differences between groups in RCMAS scores.

According to chi-square analysis and the cutoff score of 19, the study group had a significantly higher prevalence of anxiety (54.5%, CI95 40.07–68.3%) than the controls (4.8%, CI95 1.66–13.29%) ($\chi^2 = 33.496$, $p < 0.001$). The Fischer exact test indicated a significant difference between groups ($p < 0.001$).

Hypothesis 1. *Differences between the study groups in sensory modulation.*

3.3. Subjective Measure—The AASP Self-Report

According to the MANCOVA test, a significantly higher prevalence of SMDs was found among the study group as manifested in the four AASP sensory profiles: adolescents with anxiety disorders had significantly greater sensory sensitivity and avoidance, lower ability to register sensory input, and lower tendency to seek sensory input. The effect size of the observed variables of SMDs was the partial eta square η_p^2 that ranged from 0.145 to 0.278, indicating large effect sizes (see Table 5, Figure 4). As presented in Table 6, according to chi-square analysis, this was also manifested in higher prevalence of abnormal SMDs (based on AASP normal/abnormal ranges) among the study group, which showed greater sensory sensitivity ($\chi^2 = 26.025$, $p < 0.001$), higher difficulty to register sensory input ($\chi^2 = 21.328$, $p < 0.001$), greater sensation avoidance ($\chi^2 = 29.457$, $p < 0.001$), and lower tendency to seek for sensory input ($\chi^2 = 19.082$, $p < 0.001$). The Fischer exact test indicated a significant difference between groups ($p < 0.001$).

Table 5. Comparing AASP scores between groups according to the MANCOVA test.

	Study Group (n = 43) Mean + SE	Control Group (n = 62) Mean + SE	$F_{(1,105)}$	η_p^2	p	95% CIs Lower Bound	95% CIs Upper Bound
Sensory sensitivity	46.311 ± 1.599	32.832 ± 1.312	39.266	0.278	<0.001 ***	9.212	17.745
Low registration	39.388 ± 1.473	31.150 ± 1.208	17.282	0.145	<0.001 ***	4.307	12.168
Sensory seeking	41.787 ± 1.101	49.406 ± 0.903	26.483	0.206	<0.001 ***	−10.556	−4.682
Sensation avoidance	40.680 ± 1.408	30.190 ± 1.154	30.701	0.231	<0.001 ***	6.735	14.245

Note: SE = standard error; η_p^2 (partial η^2) = effect size; *** $p < 0.001$.

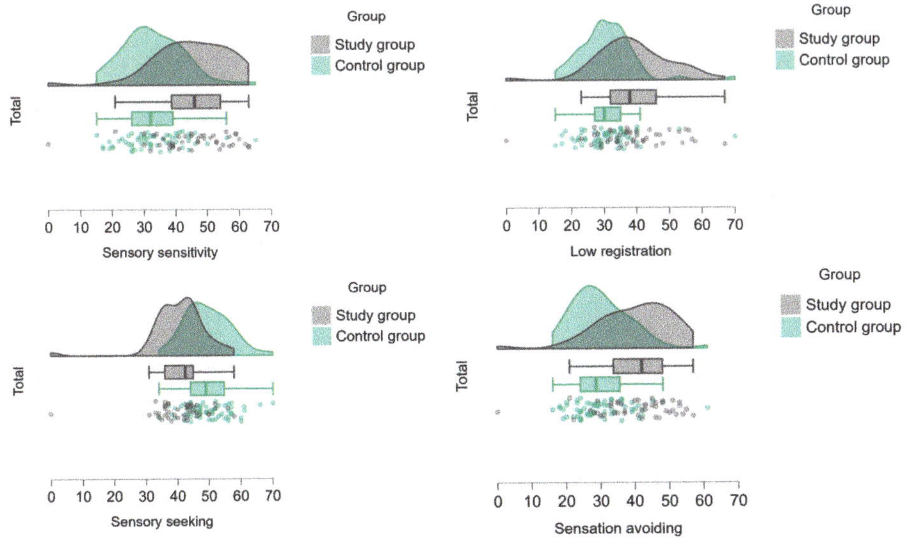

Figure 4. Differences between groups in AASP scores.

Table 6. Prevalence of SMDs among the study group as compared to the controls according to the AASP, according to chi-square analysis.

Sensory Model	Group	Study Group n	%	95% CIs Lower Bound	Upper Bound	Control Group n	%	95% CIs Lower Bound	Upper Bound	χ^2 (2)	p	Fischer
Sensory sensitivity	Below norm	1	2.3	0.004	0.120	12	19.4	0.114	0.308			<0.001 ***
	At norm	13	30.2	0.186	0.451	38	61.3	0.488	0.724	26.025	<0.001 ***	
	Above norm	29	67.4	0.525	0.795	12	19.4	0.114	0.308			
Low registration	Below norm	2	4.7	0.012	0.154	15	24.2	0.152	0.361			
	At norm	25	58.1	0.433	0.716	44	71	0.587	0.807	21.328	<0.001 ***	
	Above norm	16	37.2	0.243	0.521	3	4.8	0.016	0.132			
Sensory seeking	Below norm	19	44.2	0.304	0.589	6	9.7	0.045	0.195			
	At norm	24	55.8	0.411	0.695	50	80.6	0.691	0.885	19.082	<0.001 ***	
	Above norm	0	0	0	0.082	6	9.7	0.045	0.195			
Sensation avoidance	Below norm	2	4.7	0.012	0.154	20	32.3	0.219	0.446			
	At norm	16	37.2	0.243	0.521	35	56.4	0.440	0.680	29.457	<0.001 ***	
	Above norm	25	58.1	0.433	0.716	7	11.3	0.055	0.215			

*** $p < 0.001$.

3.4. Objective Measures–EDA and Heartrate

According to the MANCOVA test, greater reactivity to auditory startles was found in the EDA of the study group as compared to the controls. In response to the first stimulus, the study group had a significantly shorter reaction time and higher amplitude height. When referring to session 2 (the startles session) and 3 (the relaxation session), the study group had a significantly higher height of the maximum amplitude, reflecting higher arousability and difficulties in adapting and relaxing.

According to the ANCOVA test, the study group also showed higher heart rate than the control group at the beginning and the end of the EDA test. The effect size of all the observed variables was the partial eta square $\eta_p{}^2$ that ranged from 0.048 to 0.386, indicating small to large effect sizes (see Table 7, Figures 5 and 6). Nevertheless, no significant differences were found between the study groups in the heart rate delta (starting point and ending of the EDA procedure).

Table 7. Comparing EDA and heart rate between groups according to MANCOVA and ANCOVA tests.

			Study Group (n = 37) Mean + SE	Control Group (n = 60) Mean + SE	$F_{(1,97)}$	$\eta_p{}^2$	p	95% CIs Lower Bound	Upper Bound
Phase 2–first startle	Arousability	Response time to 1st stimulus	72.043 ± 0.268	74.757 ± 0.206	59.206	0.386	<0.001 ***	−3.413	−2.013
		Amplitude height—1st stimulus	4.776 ± 0.453	3.372 ± 0.348	5.553	0.056	0.021 *	0.221	2.588
			Study Group (n = 43) Mean + SE	**Control Group (n = 62) Mean + SE**	**$F_{(1,105)}$**	**$\eta_p{}^2$**			
Phase 2–startles session	Arousability	Time to get to maximum amplitude	147.979 ± 8.909	161.840 ± 7.307	1.338	0.013	0.250	130.304	165.647
		Height of the maximum amplitude	5.952 ± 0.490	4.062 ± 0.402	8.213	0.075	0.005 **	4.980	6.925
	Adaptation	Time to get to minimum amplitude	156.328 ± 9.409	151.482 ± 7.717	0.147	0.001	0.703	137.665	174.991
		Height of the minimum amplitude	3.098 ± 0.314	2.696 ± 0.258	0.903	0.009	0.344	2.475	3.720

Table 7. *Cont.*

			Study Group (n = 37) Mean + SE	Control Group (n = 60) Mean + SE	$F_{(1,97)}$	η_p^2	p	95% CIs Lower Bound	Upper Bound
Phase 3–relaxation session	Arousability	Time to get to maximum amplitude	268.962 ± 3.567	276.430 ± 2.925	2.423	0.023	0.123	261.887	276.037
		Height of the maximum amplitude	5.303 ± 0.491	3.718 ± 0.403	5.762	0.053	0.018 *	4.330	6.277
	Adaptation	Time to get to minimum amplitude	282.181 ± 3.60	279.326 ± 2.953	0.348	0.003	0.557	275.040	289.323
		Height of the minimum amplitude	3.406 ± 0.396	3.131 ± 0.325	0.260	0.003	0.611	2.616	4.189
Heartrate (phases 1 and 3)	Arousability	Pulse rate at the beginning of the EDA test	90.810 ± 2.372	81.683 ± 1.972	8.126	0.073	0.005 **		
	Adaptation	Pulse rate at the end of the EDA test	88.053 ± 2.11	81.543 ± 1.757	5.205	0.048	0.025 *		

Note: SE = standard error; η_p^2 (partial η^2) = effect size; EDA = Electro-dermal activity; * $p < 0.05$; ** $p < 0.01$; *** $p \leq 0.001$.

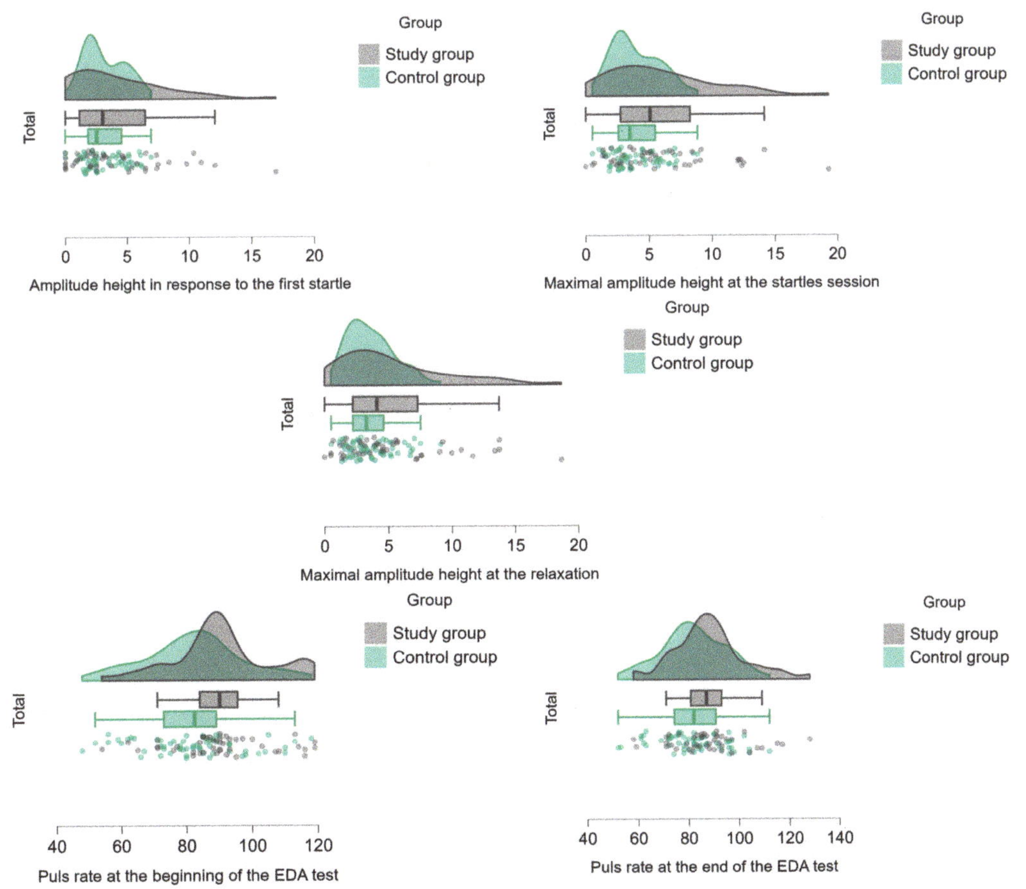

Figure 5. Differences between groups in EDA and pulse rate scores.

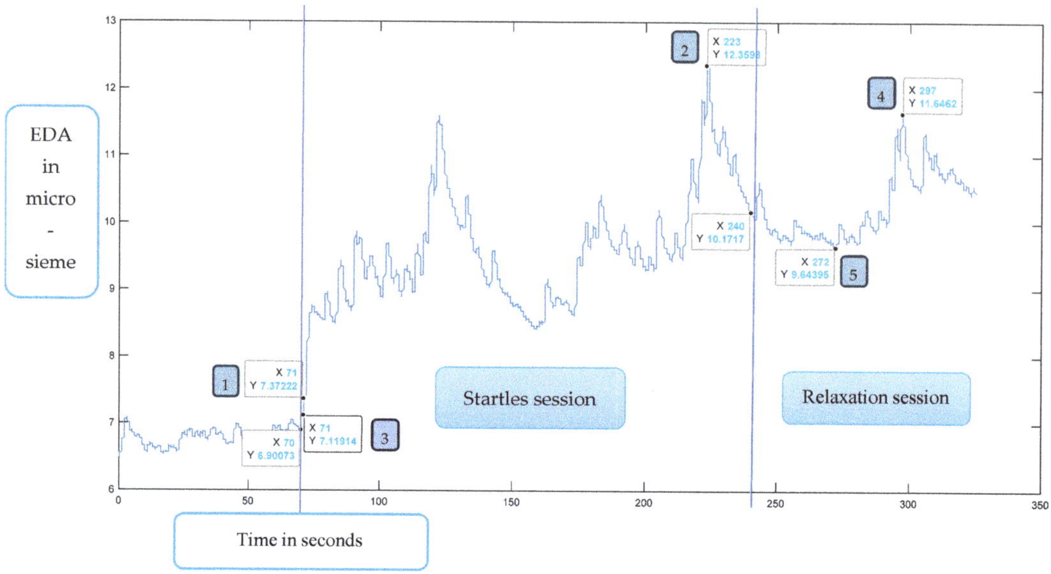

Figure 6. EDA recording of a study group participant. Note: 1 = reaction time and amplitude height in response to the first startle; 2 = time and height of the maximum amplitude during the 3 min of auditory startles; 3 = the time and height of minimal amplitude during the 3 min of auditory startles session; 4 = time and height of maximum amplitude at the time of relaxation session; 5 = time and height of minimum amplitude at the time of relaxation session.

Hypothesis 2. *Correlations between variables in the study group.*

3.5. The Correlations Between Subjective and Objective Measures

According to the Pearson correlation test, medium significant negative correlations were found between the AASP and the physiological manifestations—the EDA and heart rate. Lower sensory seeking correlated with higher amplitude height in response to the first startle (r = −0.327, p = 0.05), with the maximal (r = −0.415, p = 0.007) and the minimal amplitude (r = −0.458, p = 0.003) during the startles' session as well as with the maximal (r = −0.332, p = 0.034) and the minimal amplitude (r = −0.355, p = 0.023) after the startles' session—in the relaxation session. Lower sensory seeking also correlated with the time of the maximal amplitude (r = −0.340, p = 0.03) in relaxation and with elevated heart rate at the beginning of the EDA test (before starting the startles session) (r = −0.355, p = 0.034).

3.6. The Correlations Between the Subjective AASP and the Psychopathological Symptoms in Both Groups

As reported in Table 8, according to the Pearson correlation test, in children with anxiety, medium to large significant positive correlations were found between the AASP patterns' scores and anxiety psychopathological symptoms according to the RCMAS: greater sensory sensitivity and sensation avoidance significantly positively correlated with higher levels of anxiety, especially, with the physiological, worry, and over-responsivity, social concern problems, and total anxiety with large effect sizes. Lower registration of sensory input significantly positively correlated with the RCMAS physiological scale and total anxiety score with medium effect sizes.

Table 8. The correlations between sensory modulation according to the AASP and psychopathological symptoms according to the RCMAS and the CBCL in the study group according to the Pearson correlation test.

		Study Group (n = 43)				**Control Group (n = 62)**			
		Sensory Sensitivity	**Low Registration**	**Sensory Seeking**	**Sensation Avoidance**	**Sensory Sensitivity**	**Low Registration**	**Sensory Seeking**	**Sensation Avoidance**
RCMAS	Physiological	0.531 ***	0.471 **	0.48	0.721 ***	0.581 ***	0.415 ***	−0.085	0.427 ***
	Worry and over-responsivity	0.482 ***	0.278	0.052	0.570 ***	0.679 ***	0.462 ***	−0.149	0.520 ***
	Social concern and concentration	0.544 ***	0.194	0.258	0.536 ***	0.523 ***	0.442 ***	−0.144	0.350 **
	Lie	−0.047	−0.090	−0.044	−0.124	−0.250	−0.180	0.192	−0.118
	Total anxiety score	0.621 ***	0.392 **	0.123	0.731 ***	0.715 ***	0.533 ***	−0.165	0.513 ***
		Study Group (n = 42)				**Control Group (n= 60)**			
CBCL	Anxious/Depressed	0.147	−0.025	0.132	0.212	0.351 **	0.273 *	−0.066	0.344 **
	Withdrawn/Depressed	0.277	0.226	−0.044	0.482 ***	0.073	0.158	−0.069	0.159
	Somatic complaints	0.236	0.155	0.196	0.309 *	0.289 *	0.227	0.020	0.223
	Social problems	0.074	0.103	0.070	0.167	0.168	0.225	0.066	0.156
	Thought problems	0.150	0.118	0.169	0.167	0.042	−0.006	−0.031	−0.008
	Attention problems	−0.107	0.209	0.125	0.038	0.034	0.101	−0.102	0.083
	Rule-breaking	−0.095	−0.272	0.169	−0.053	−0.030	0.186	0.184	0.090
	Aggressive behavior	−0.086	−0.012	0.089	0.112	−0.050	0.001	0.118	0.059
	Other conditions	0.167	0.029	0.048	0.170	−0.116	−0.067	0.213	−0.089
	Internalization	0.156	0.058	0.074	0.325 *	0.307	0.240	−0.071	0.290 *
	Externalization	−0.041	−0.132	0.099	0.007	−0.035	0.043	0.155	0.105
	Total problems	0.088	0.053	0.124	0.206	0.074	0.091	0.081	0.148

Note: * $p \leq 0.05$; ** $p \leq 0.01$; *** $p \leq 0.001$.

In the control group, medium to large significant positive correlations were found between the AASP patterns' scores and anxiety psychopathological symptoms according to the RCMAS: greater sensory sensitivity, low registration and sensation avoidance significantly positively correlated with higher levels of anxiety, especially, with the physiological, worry and over-responsivity, social concern problems, and total anxiety.

Medium significant positive correlations were found between AASP scales and CBCL scales. In the study group: sensation avoidance significantly positively correlated with emotional psychopathological symptoms of withdrawn/depressed, somatic complaints, and internalization. AASP sensation avoidance correlated with worse CBCL psychopathological symptoms of withdrawal/depression, somatic complaints, and internalization. In the control group, greater sensory sensitivity significantly positively correlated with higher anxious/depressed symptoms and somatic complaints; greater low registration significantly positively correlated with higher anxious/depressed symptoms; greater sensation avoidance significantly positively correlated with higher anxious/depressed symptoms and internalization. (See Table 8).

4. Discussion

The present study aimed to profile SMDs in adolescents with anxiety disorders and examine their relation to psychopathology. This study used subjective self-reports to reflect SMDs in daily life scenarios from the individual's own voice, and objective measures to reflect underlying mechanisms that may explain behavioral responses and functional restrictions.

The first hypothesis was confirmed: SMDs were more prevalent among the study group. The AASP subjective self-report found that adolescents with anxiety disorders had higher sensory sensitivity and avoidance, lower ability to register sensory input, and lower tendency to seek sensory input than the controls. A recent meta and retrospective analysis of people with various psychiatric conditions found that based on AASP, SMDs were prevalent and expressed in elevated sensory sensitivity, sensation avoidance, low registration, and a lower tendency to seek sensory input [30,31]. Our results confirm earlier findings according to which SMDs are prevalent among children and adolescents with anxiety disorders [32].

Previous reports which also used both self-reports and objective measures (such as EDA, and functional neuroimaging), were performed on various age groups—from early childhood to adulthood but referred to multiple psychiatric disorders. Tolin et al. (2021) who focused on adults with anxiety disorders used EDA and heart rate recordings in response to auditory startles and found sympathetic hyper-arousability according to both measures. They also found lower autonomic recovery after removing the auditory stressor [18].

Our study focused on anxiety in adolescence and highlighted how important it is to elaborate on the knowledge regarding SMDs, their relation to anxiety severity, and their functional implications in this critical developmental stage. Moreover, our study emphasized the sensitivity of the AASP in demonstrating unmodulated perception of sensory stimuli among adolescents with anxiety disorders. Support for that was found by the objective measures (EDA and heart rate) which depicted higher reactivity and lower adaptation to sensory input. Even though most adolescents in the study group were under medications, they still exhibited an elevated physiological hyper-arousability—in response to the first stressor auditory startle, during the startles' session (in which adaptation should have been found, in response to the repeated auditory stimuli), and even during the third session of relaxation, emphasizing the lingering effect of sensory stimuli. This information may shed light on daily function obstacles that these adolescents experience. Therefore, hyper-arousability should be evaluated in adolescents with anxiety disorders in order to deeply understand its consequences on patients' functioning and pathogenesis severity.

With respect to the second hypothesis, SMDs (both sensory sensitivity and sensation avoidance), significantly correlated with anxiety psychopathology severity as manifested in RCMAS results (higher levels of worry and over-responsivity, social concern, and concentration problems) in both study groups. In the study group, the relations with anxiety psychopathology but not with other psychopathological symptoms of the CBCL may be related to the fact that this measure includes behavioral and externalizing symptoms not related to anxiety. Moreover, CBCL was measured according to parents' reports, while the RCMAS was according to the adolescents' reports themselves. The relation between SMDs and psychopathology severity was found in studies on other clinical populations. For example, MacLennan, Roach, and Tavassoli (2020) studied 41 autistic children aged between 3 and 14 years using parent and self-reported measures and found that greater sensory hyperreactivity correlated with elevated phobia-related symptoms, total anxiety, and separation anxiety [33]. Interestingly, in the present study, patterns related to sensory hypersensitivity correlated with worse anxiety psychopathology, while sensory seeking correlated with a better clinical profile and more adaptive reactivity and modulation to sensory input. Our results are consistent with prior research on patients with psychiatric disorders indicating that sensory seeking correlated with a better clinical profile and a better quality of life and, therefore, might presumably serve as a protective pattern [34].

The results of the present study support the previous literature highlighting the role of SMDs in anxiety development [35] and psychopathology [6,7]. Studies found that SMDs in adolescents with anxiety disorders are related to nonadaptive behavior [36], worse social functioning [37], and sleep problems [38]. Therefore, SMDs in anxiety disorders should receive clinical and scientific attention [31]. Detailed evaluation of sensory modulation should be performed based on self-reports and objective physiological measures. Non-invasive physiological indicators of autonomic activity [39] such as EDA offer an excellent opportunity to assess and monitor the physiological alterations under different conditions of anxiety [40,41]. Practically, these measures may exhibit clinical utility for aiding clinicians in monitoring the progress of treatment [42,43]. In research, these measures may help in conceptualizing biopsychological processes that can be targeted to optimize care [44]. This detailed information may focus intervention according to the individual unique sensory profiles, unique pathogenesis, and real-life context [34].

In therapy, sensory-based intervention strategies may be relevant for adolescents with anxiety disorders [6,45]. Therapists should elevate the awareness of adolescents, parents, teachers, and health providers to the involvement of SMDs in the pathogenesis of anxiety and on daily function; Applying sensory stimuli that fit the individual's sensory profile may assist in modulating behavior and function. For example, weighted blankets, Yuckee balls, and wraps may enhance the perception of sensory input to low registrators. For hyper-sensitive individuals, relaxation techniques using sensory-based strategies (such as massage) may be relevant [45]. Sensory modulation intervention may be combined with cognitive behavioral therapy (CBT) and provide strategies to cope with challenging sensory situations in daily life [45].

5. Conclusions

SMDs may be prevalent in adolescents with anxiety disorders and explain behavioral hyper-arousablity and functional deficits. SMDs may be related to psychopathological symptoms of anxiety. Since sensory modulation plays a crucial role in the way we perceive the world, connect with others, and function, SMDs should be evaluated in adolescents with anxiety disorders. Evaluations should include self-reports and objective physiological measurements. This may improve intervention outcomes, decrease in the severity of the psychopathological symptoms, and improve function, development, and well-being.

6. Research Limitations

The findings of the present study should be interpreted in light of the following limitations: the study groups varied in certain socio-demographic parameters. The study group included various anxiety disorders with different coexistence psychiatric disorders. Therefore, it is recommended to conduct future studies on larger samples with similar socio-demographic characteristics and specific anxiety disorders to enable the generalizability of the results.

Regarding discrepancies identified between the RCMAS and CBCL questionnaires' correlational results with the AASP, these should be assessed by principal component analysis (PCA) or other multivariate analysis in future studies, using a larger sample.

Finally, most participants in the study group were under psychiatric medication (such as antidepressants), future studies should consider the possible impacts of medications on the results.

Author Contributions: Conceptualization G.H., A.A.-M., H.J., R.A.-T. and B.E.-Y.; methodology G.H. and B.E.-Y.; formal analysis G.H., H.J., R.A.-T. and B.E.-Y.; data curation, G.H., A.A.-M., H.J. and R.A.-T.; writing—original draft preparation, G.H. and B.E.-Y.; writing—review and editing G.H. and B.E.-Y.; supervision, B.E.-Y. and A.A.-M.; project administration, G.H., A.A.-M., H.J. and R.A.-T. All authors have read and agreed to the published version of the manuscript.

Funding: This research received no external funding.

Institutional Review Board Statement: The study was conducted in accordance with the Declaration of Helsinki [BLINDED] (approval code: 0045-22-EMC, approved on 31 October 2022) and the

Institutional Review Board of the [BLINDED] (approval code: 333/18, approved on 17 September 2018; an extension- 333/18-1 was approved on 18 December 2022).

Informed Consent Statement: Informed consent was obtained from all subjects involved in the study.

Data Availability Statement: Data are not publicly available due to privacy and ethical restrictions.

Conflicts of Interest: The authors declare no conflicts of interest.

References

1. Racine, N.; McArthur, B.A.; Cooke, J.E.; Eirich, R.; Zhu, J.; Madigan, S. Global prevalence of depressive and anxiety symptoms in children and adolescents during COVID-19: A meta-analysis. *JAMA Pediatr.* **2021**, *175*, 1142–1150. [CrossRef]
2. American Psychiatric Association. *Diagnostic and Statistical Manual of Mental Disorders: DSM-5*; American Psychiatric Association: Washington, DC, USA, 2013; Volume 5.
3. D'Agati, E.; Curatolo, P.; Mazzone, L. Comorbidity between ADHD and anxiety disorders across the lifespan. *Int. J. Psychiatry* **2019**, *23*, 238–244. [CrossRef]
4. Kessler, R.C.; Chiu, W.T.; Demler, O.; Walters, E.E. Prevalence, severity, and comorbidity of 12-month DSM-IV disorders in the National Comorbidity Survey Replication. *Arch. Gen. Psychiatry* **2005**, *62*, 617–627. [CrossRef]
5. Ramsawh, H.J.; Chavira, D.A.; Stein, M.B. Burden of anxiety disorders in pediatric medical settings: Prevalence, phenomenology, and a research agenda. *Arch. Pediatr. Adolesc. Med.* **2010**, *164*, 965–972. [CrossRef]
6. Wallis, K.; Sutton, D.; Bassett, S. Sensory modulation for people with anxiety in a community mental health setting. *Occup. Ther. Ment. Health* **2018**, *34*, 122–137. [CrossRef]
7. MacLennan, K.; Rossow, T.; Tavassoli, T. The relationship between sensory reactivity, intolerance of uncertainty and anxiety subtypes in preschool-age autistic children. *Autism* **2021**, *25*, 2305–2316. [CrossRef] [PubMed]
8. Rodriguez, M.; Kross, E. Sensory emotion regulation. *Trends Cogn. Sci.* **2023**, *27*, 379–390. [CrossRef]
9. Engel-Yeger, B.; Dunn, W. The relationship between sensory processing difficulties and anxiety level of healthy adults. *Br. J. Occup. Ther.* **2011**, *74*, 210–216. [CrossRef]
10. Serafini, G.; Engel-Yeger, B.; Vazquez, G.H.; Pompili, M.; Amore, M. Sensory processing disorders, duration of current episode, and severity of side effects in major affective and anxiety disorders. *Eur. Psychiatr.* **2016**, *33*, s224–s225. [CrossRef]
11. Dunn, W. The impact of sensory processing abilities on the daily lives of young children and their families: A conceptual model. *Infants Young Child.* **1997**, *9*, 23–35. [CrossRef]
12. Dunn, W. *Sensory Profile 2*; Psych Corporation: Bloomington, MN, USA, 2014.
13. Dunn, W. *Sensory Profile: User's Manual*; Psychological Corporation: San Antonio, TX, USA, 1999.
14. Mubarak, A.; Cyr, C.; St-André, M.; Paquette, D.; Emond-Nakamura, M.; Boisjoly, L.; Palardy, S.; Adin, S.; Stikarovska, I. Child attachment and sensory regulation in psychiatric clinic-referred preschoolers. *Clin. Child Psychol. Psychiatry* **2017**, *22*, 572–587. [CrossRef] [PubMed]
15. Lane, S.J.; Reynolds, S.; Thacker, L. Sensory over-responsivity and ADHD: Differentiating using electrodermal responses, cortisol, and anxiety. *Front. Integr. Neurosci.* **2010**, *4*, 8. [CrossRef] [PubMed]
16. McLean, M.A.; Niknafs, N.; Scoten, O.C.; Chau, C.M.; MacKay, M.; Weinberg, J.; Synnes, A.; Miller, S.P.; Grunau, R.E. Sensory processing and cortisol at age 4 years: Procedural pain-related stress in children born very preterm. *Dev. Psychobiol.* **2021**, *63*, 915–930. [CrossRef] [PubMed]
17. Kushki, A.; Drumm, E.; Pla Mobarak, M.; Tanel, N.; Dupuis, A.; Chau, T.; Anagnostou, E. Investigating the autonomic nervous system response to anxiety in children with autism spectrum disorders. *PLoS ONE* **2013**, *8*, e59730. [CrossRef]
18. Tolin, D.F.; Lee, E.; Levy, H.C.; Das, A.; Mammo, L.; Katz, B.W.; Diefenbach, G.J. Psychophysiological assessment of stress reactivity and recovery in anxiety disorders. *J. Anxiety Disord.* **2021**, *82*, 102426. [CrossRef] [PubMed]
19. Pruneti, C.A.; Lento, R.M.; Fante, C.; Carrozzo, E.; Fontana, F. Autonomic arousal and differential diagnosis in clinical psychology and psychopathology. *J. Psychopathol.* **2010**, *16*, 43–52.
20. Kotianova, A.; Kotian, M.; Slepecky, M.; Chupacova, M.; Prasko, J.; Tonhajzerova, I. The differences between patients with panic disorder and healthy controls in psychophysiological stress profile. *Neuropsychiatr. Dis. Treat.* **2018**, *14*, 435–441. [CrossRef] [PubMed]
21. Dieleman, G.C.; Huizink, A.C.; Tulen, J.H.; Utens, E.M.; Creemers, H.E.; van der Ende, J.; Verhulst, F.C. Alterations in HPA-axis and autonomic nervous system functioning in childhood anxiety disorders point to a chronic stress hypothesis. *Psychoneuroendocrinology* **2015**, *51*, 135–150. [CrossRef] [PubMed]
22. Engel-Yeger, B. The involvement of altered sensory modulation in neurological conditions and its relevance to neuro-rehabilitation: A narrative literature review. *Disabil. Rehabil.* **2021**, *43*, 2511–2520. [CrossRef] [PubMed]

23. Achenbach, T.M.; Rescorla, L. *Manual for the ASEBA School-Age Forms and Profiles*; University of Vermont Research Center for Children, Youth, & Families: Burlington, VT, USA, 2001.
24. Reynolds, C.R.; Richmond, B.O. *Revised Children's Manifest Anxiety Scale. RCMAS Manual*; Western Psychological Services: Los Angeles, CA, USA, 1985.
25. Reynolds, C.R.; Harding, R.E. Outcome in two large sample studies of factorial similarity under six methods of comparison. *Educ. Psychol. Meas.* **1983**, *43*, 723–728. [CrossRef]
26. Brown, C.; Dunn, W. *Adolescent-Adult Sensory Profile: User's Manual*; Therapy Skill Builders: San Antonio, TX, USA, 2002.
27. Serafini, G.; Engel-Yeger, B.; Vazquez, G.H.; Pompili, M.; Amore, M. Sensory Processing Disorders are Associated with Duration of Current Episode and Severity of Side Effects. *Psychiatry Investig.* **2017**, *14*, 51–57. [CrossRef] [PubMed]
28. Fowles, D.C. The eccrine system and electrodermal activity. *Psychophysiol. Syst. Process. Appl.* **1986**, *1*, 51–96.
29. Parma, V.; Cellini, N.; Guy, L.; McVey, A.J.; Rump, K.; Worley, J.; Maddox, B.; Bush, J.; Bennett, A.; Franklin, M.; et al. Profiles of autonomic activity in autism spectrum disorder with and without anxiety. *J. Autism Dev. Disord.* **2021**, *51*, 4459–4470. [CrossRef] [PubMed]
30. Brown, C.; Karim, R.; Steuter, M. Retrospective analysis of studies examining sensory processing preferences in people with a psychiatric condition. *Am. J. Occup. Ther.* **2020**, *74*, 7404205130p1–7404205130p11. [CrossRef] [PubMed]
31. Van den Boogert, F.; Klein, K.; Spaan, P.; Sizoo, B.; Bouman, Y.H.; Hoogendijk, W.J.; Roza, S.J. Sensory processing difficulties in psychiatric disorders: A meta-analysis. *J. Psychiatr. Res.* **2022**, *151*, 173–180. [CrossRef]
32. Houghton, D.C.; Stein, D.J.; Cortese, B.M. Exteroceptive sensory abnormalities in childhood and adolescent anxiety and obsessive-compulsive disorder: A critical review. *J. Am. Acad. Child Adolesc. Psychiatry* **2020**, *59*, 78–87. [CrossRef]
33. MacLennan, K.; Roach, L.; Tavassoli, T. The relationship between sensory reactivity differences and anxiety subtypes in autistic children. *Autism Res.* **2020**, *13*, 785–795. [CrossRef] [PubMed]
34. Engel-Yeger, B.; Muzio, C.; Rinosi, G.; Solano, P.; Geoffroy, P.A.; Pompili, M.; Amore, M.; Serafini, G. Extreme sensory processing patterns and their relation with clinical conditions among individuals with major affective disorders. *J. Psychiatr. Res.* **2016**, *236*, 112–118. [CrossRef]
35. McMahon, K.; Anand, D.; Morris-Jones, M.; Rosenthal, M.Z. A path from childhood sensory processing disorder to anxiety disorders: The mediating role of emotion dysregulation and adult sensory processing disorder symptoms. *Front. Integr. Neurosci.* **2019**, *13*, 22. [CrossRef]
36. Mulligan, S.; Douglas, S.; Armstrong, C. Characteristics of idiopathic sensory processing disorder in young children. *Front. Integr. Neurosci.* **2021**, *15*, 647928. [CrossRef]
37. John, T.S.; Estes, A.; Begay, K.K.; Munson, J.; Reiter, M.A.; Dager, S.R.; Kleinhans, N. Characterizing social functioning in school-age children with sensory processing abnormalities. *J. Autism Dev. Disord.* **2022**, *52*, 1361–1373. [CrossRef] [PubMed]
38. Mazurek, M.O.; Petroski, G.F. Sleep problems in children with autism spectrum disorder: Examining the contributions of sensory over-responsivity and anxiety. *Sleep Med.* **2015**, *16*, 270–279. [CrossRef] [PubMed]
39. Nikolić, M.; Aktar, E.; Bögels, S.; Colonnesi, C.; De Vente, W. Bumping heart and sweaty palms: Physiological hyperarousal as a risk factor for child social anxiety. *J. Child Psychol. Psychiatry* **2018**, *59*, 119–128. [CrossRef]
40. Sebastião, R. Classification of Anxiety Based on EDA and HR. In *International Conference on IoT Technologies for HealthCare*; Springer International Publishing: Cham, Switzerland, 2020; pp. 112–123.
41. Desai, U.; Shetty, A.D. Electrodermal activity (EDA) for treatment of neurological and psychiatric disorder patients: A review. In Proceedings of the 2021 7th International Conference on Advanced Computing and Communication Systems (ICACCS), IEEE, Coimbatore, India, 19–20 March 2021; Volume 1, pp. 1424–1430.
42. McCormack, C.C.; Mennies, R.J.; Silk, J.S.; Stone, L.B. How anxious is too anxious? State and trait physiological arousal predict anxious youth's treatment response to brief cognitive behavioral therapy. *BMC Psychol.* **2020**, *8*, 48. [CrossRef] [PubMed]
43. Pop-Jordanova, N.; Pop-Jordanov, J. Electrodermal activity and stress assessment. *Prilozi* **2020**, *41*, 5–15. [CrossRef] [PubMed]
44. Christian, C.; Cash, E.; Cohen, D.A.; Trombley, C.M.; Levinson, C.A. Electrodermal activity and heart rate variability during exposure fear scripts predict trait-level and momentary social anxiety and eating-disorder symptoms in an analogue sample. *Clin. Psychol. Sci.* **2023**, *11*, 134–148. [CrossRef]
45. Mavhunga, T. The Effectiveness, Acceptability and Practicality of a CBT-Sensory Modulation Intervention for Children 4–7 Years Old Presenting with Anxiety. Ph.D. Thesis, Auckland University of Technology, Auckland, New Zealand, 2024.

children

MDPI

Article

A School-Based Intervention Program to Reduce Weight Stigma in Adolescents

Mariacarolina Vacca [1,*], Silvia Cerolini [2], Anna Zegretti [3], Andrea Zagaria [4], Edoardo Mocini [5] and Caterina Lombardo [3]

1 Department of Experimental Medicine, Sapienza University of Rome, 00185 Rome, Italy
2 Department of Social Science, Guglielmo Marconi University, 00193 Rome, Italy
3 Department of Psychology, Sapienza University of Rome, 00185 Rome, Italy
4 Department of Systems Medicine, Tor Vergata University of Rome, 00133 Rome, Italy
5 Department of Theoretical and Applied Sciences, eCampus University, 22100 Novedrate, Italy
* Correspondence: mariacarolina.vacca@uniroma1.it

Abstract: Background/Objectives: Weight-based stigma represents a pervasive psychosocial challenge affecting youth worldwide, with significant implications for mental and physical health. Although school-based interventions have been suggested as effective strategies to reduce weight bias, evidence regarding their content and efficacy in real-world educational contexts remains limited. The present study aimed to evaluate the effectiveness of a school-based intervention designed to reduce internalized weight stigma among Italian adolescents. Methods: A total of 539 secondary school students (Mage = 15.91 years; SD = 1.38; 51.5% males) from 10 schools in Italy were randomly assigned to either an intervention group (61.2%, $n = 330$) receiving the program or a minimal intervention control group (38.8%, $n = 209$). The intervention integrated psychoeducational modules and activities based on cognitive dissonance theory to address maladaptive weight-related attitudes and associated psychopathological outcomes. Pre- and post-intervention assessments evaluated internalized weight bias and body dissatisfaction in both groups. Results: A significant reduction in internalized weight bias was observed among students perceiving themselves as overweight following the intervention. Additionally, a marginally significant decrease in body dissatisfaction was reported in the intervention group from pre- to post-intervention. No comparable changes were found in the control group. Conclusions: These findings suggest that incorporating cognitive dissonance-based psychoeducational content within school-based programs may be a promising approach for reducing internalized weight stigma in adolescents. Further research is warranted to refine intervention strategies, explore the mechanisms underlying the observed effects, and assess the scalability and long-term impact of such initiatives in school settings.

Keywords: weight stigma; adolescents; secondary school; intervention

1. Introduction

The past decade has seen a renewed importance in the concept of weight-based stigma [1], defined as negative weight-related attitudes toward people who present overweight (OW) or obesity (OB), expressed through negative stereotyping and/or discrimination, resulting in social denigration and prejudice [2,3]. Weight stigma is an overriding public health concern, particularly considering the steadily increasing prevalence of OB and OW and their health and psychological complications [4]. The worldwide prevalence rates of overweight and obesity amount to one-third of the global population, and current trends continue to rise dramatically [5], with an estimated reduction in life expectancy of 5–20 years depending on the severity of the condition [6].

In Italy, over ten percent of the adult population is affected by OB, and over one third presents OW [7]. Alarming rates of OW/OB have been observed in youth, more specifically among children [8] (30%) and adolescents (14.6%) [9], reaching the highest prevalence in Europe [10]. It has been reported that 40–50% of adults with OB/OW report perceiving negative weight-related biases, with 20% of them experiencing extremely high weight bias stigma [11]. Although obesity is associated with health risks, evidence indicates that individuals can maintain good health at higher body weights, highlighting the importance of body diversity and the need to reduce weight stigma [12].

Weight stigma may be particularly detrimental and treacherous when experienced during adolescence, a developmental period (i.e., 10–19 years) [13] accompanied by numerous body-related changes [14] and a contemporary emphasis on physical appearance [15]. Adolescence is also a critical period for identity development, during which body image and body satisfaction become closely linked to self-concept and social identity formation [16]. During this transitional period, physiological, social, and psychological changes make adolescents particularly vulnerable to developing a negative body image (i.e., an internalized view of one's appearance) [17] and body dissatisfaction [18]. Friends, family, and peers are the most common sources of weight-based stigma in youth [19]. Adolescents are extremely vulnerable to weight-related judgment and stigma, especially from peers [20] and primarily in the school context [21]. The increasing awareness of physical appearances makes teenagers more prone to compare their bodies with those of peers and with the media's portrayal of the "ideal" body [15,22]. Social media platforms can exacerbate weight stigma by increasing exposure to appearance-focused content and peer comparisons, potentially worsening body dissatisfaction and internalized stigma [23]. Youth anti-fat attitudes can result in different behavioral outcomes against victims, such as verbal teasing, bullying, and social exclusion [24]. Weight bias could also be negatively internalized in adolescents with OW or OB (i.e., weight self-stigma) [25] who attribute weight-based negative stereotypes to themselves, thus exacerbating the risk of social marginalization [26].

Recent estimates indicate that from 13 to 32% of adolescents report that they have been discriminated against by peers due to body weight [27]. The pernicious nature of weight stigma in youngsters has been supported by evidence suggesting that stigma exerts negative long-term effects on adolescent mental health (e.g., poorer social and emotional adjustment) [28]. Experiencing weight stigma is associated with maladaptive outcomes such as unhealthy eating, exercise and healthcare avoidance, and psychosocial difficulties including low self-esteem and social isolation [29–31]. Furthermore, weight stigma has been associated with adolescent anxiety, depression, low self-esteem, and overall poor health-related quality of life [28,32]. Longitudinal research has shown that weight-based teasing predicts dysfunctional eating and adverse weight outcomes after 15 years [3]. Evidence also demonstrated that using stigma as a weight-loss motivator is a societal misconception [33].

Understanding the widespread nature of weight stigma has encouraged innovative interventions aimed at reducing weight-related teasing among adolescents [34]. Intervention studies aimed at reducing weight stigma in youth populations remain limited and methodologically heterogeneous. Several studies have proposed education-focused interventions providing information about the multifactorial nature of obesity and its socio-cultural origins [35,36]. Other prejudice-reduction interventions promote acceptance and empathy in youngsters [37], based on the evidence that increased empathy discouraged stereotypes of stigmatized groups by reducing blame or controllability attributions and improving intergroup attitudes [38]. Results on the efficacy of weight bias intervention programs remain mixed. For example, in one eight-week prevention program with school-aged adolescents that included lectures related to weight bias, weight-related victimization (i.e., repeated exposure to negative actions or comments by others intended to harm an individual based on body weight) [39], and weight-biased attitudes significantly decreased from pre- to post-test [40]. Other studies using similar approaches did not detect changes in weight-based stereotypes [41]. Evidence suggests that school-based, multicomponent interventions incorporating psychoeducation, critical reflection on sociocultural body ideals, and promotion of inclusive peer climates may attenuate weight-biased attitudes among adolescents [20,42]. However, there is still a paucity of rigorously controlled, large-scale trials evaluating the real-world effectiveness of such programs within routine educational settings [43].

The present study was designed to address this gap by implementing and assessing an intervention targeting weight-based stigma among adolescents attending ten secondary schools in a metropolitan Italian context. The intervention comprised psychoeducational modules and cognitive dissonance-based activities, which are aimed at fostering critical awareness of weight-related stereotypes and promoting inclusive classroom norms. Importantly, the study employed a controlled design, administering pre- and post-intervention assessments to both intervention and minimal intervention control (MIC) groups using validated self-report measures. More specifically, different aspects related to weight stigma were examined. Adolescents who experience weight stigma often report heightened discrepancy between actual and ideal body image, resulting from the internalisation of societal thin-ideal standards [44]. Stigmatising experiences can exacerbate eating disturbances such as dietary restraint, binge eating, and compensatory behaviors, which may serve as attempts to conform to sociocultural body ideals or to alleviate weight-related distress [45]. Weight stigma can manifest in teasing and exclusionary behaviors toward peers with higher body weight, particularly in school contexts [46]. At the same time, adolescents experiencing internalized weight stigma are particularly vulnerable to weight-based victimization [29]

By simultaneously measuring weight stigma, eating disorder symptomatology, body dissatisfaction, and bullying involvement, this study adopts a comprehensive approach to evaluating the psychosocial outcomes of an anti-stigma intervention.

2. Materials and Methods

2.1. Participants

Adolescents were recruited from lower and upper secondary schools in the urban area of Rome. Schools were identified through prior collaborations with the research team and willingness to participate, ensuring representation of both urban and suburban contexts. Twelve schools were contacted by telephone and provided with a brief description of the project aims prior to being invited to participate in the study. Ten schools agreed to participate, and formal permission to conduct the research was obtained from school principals. Two weeks before data collection, parental consent forms and student assent forms were distributed to all students in the selected classrooms. Data collection was scheduled in agreement with individual teachers and was conducted only after written informed consent from both parents/legal caregivers and students was obtained. All the school classes were randomly assigned either to the intervention group, which participated in four structured sessions, or to a MIC that received only an introductory presentation of the project along with a brief session focused on raising awareness about weight stigma and weight-related bullying. The class-level randomization resulted in uneven group sizes (around 2:3), as the number of participants allocated in each condition was influenced by pre-existing differences in class enrollment.

To ensure participant anonymity, an alphanumeric code was assigned to each student for identification purposes prior to the administration of paper-and-pencil questionnaires, which took place during regular classroom hours. All students were invited to participate in the research without restrictions, and no compensation was provided for participation. The study was approved by the Institutional Review Board of the Department of Psychology at Sapienza University of Rome (date: 27 May 2022; prot number 0001069). The data for this study were collected with the support of a research grant from the Italian Department for Family Policies under the Presidency of the Council of Ministers.

2.2. Instruments

All students were asked to complete a series of self-reported questionnaires using their mobile phone/PC/tablet in group sessions with researchers on two occasions: before (T1) and after (T2) the intervention. The first assessment (T1) was used explore the cross-sectional relationships between a) weight stigma and aspects potentially implicated in negative attitudes toward individuals with and without OW/OB (as previously presented by Cerolini et al., 2024) [47], and between, b) bullying victimization and mental health outcomes (as previously presented by Vacca et al., 2024) [48]. During the second assessment (T2), the same set of questionnaires was administered again after the completion of the program to assess the intervention efficacy.

Socio-demographic information. Participants were required to indicate their age, sex, height, and weight.

Weight-related stigma and stereotypes. The Weight Bias Internalization Scale (WBIS) [49], in Italian adaptation form [50], was used for measuring the degree to which individuals accept and internalize negative weight-related stereotypes, attitudes, and commentary. The WBIS was specifically developed to assess the internalization of weight bias among individuals who perceive themself as being in a condition of overweight and obesity. Therefore, it was administered only to participants who classified themself as falling within the categories as reported in the first item (i.e., "Do you identify as a person with overweight or obesity?"). Participants responded to each item using a 7-point scale ranging from 1 (strongly disagree) to 7 (strongly agree). Higher scores indicate greater internalization of weight bias. Cronbach's alpha in the present study was 0.904, showing excellent internal validity.

The Attitudes Toward Obese Persons Scale (ATOP) [51] in Italian adaptation form [52] was used to assess perceptions and attitudes about people with obesity. The scale consists of 16 items rated on a six-point Likert scale, ranging from −3 (strongly disagree) to 3 (strongly agree). Examples of proposed items include: "People with obesity should not expect to lead normal lives" or "People with obesity are generally sociable". As stated by the original authors of the ATOP [51–53], lower scores indicated more negative attitudes. Cronbach's alpha in the present sample was 0.765.

Eating Symptomatology. The Disordered Eating Questionnaire (DEQ) [54,55] evaluates eating difficulties and demonstrates high internal consistency. It consists of three sections: Section A includes 18 items focused on dysfunctional eating behaviors and four items concerning purging behaviors, rated on a 6-point frequency scale (0 = never; 6 = several times a day). Section B focuses on shape and weight concerns, rated on a 7-point Likert scale (0 = not at all; 7 = completely), and the last section (C) asks for the age, height, and weight of participants. All questions are presented concerning the period of the last three months. The questionnaire demonstrated good reliability in the validation sample ($\alpha = 0.90$) [54]. Cronbach's alpha in the present study was 0.935.

Body Dissatisfaction. The Contour Drawing Rating Scale (CDRS) [56] consists of nine gender-specific silhouettes with multiple body sizes ranging from the thinnest to the largest.

Participants are asked to select the silhouette that best represents their current body size and the one that reflects their ideal body size. The discrepancy between these two choices was calculated, with negative ratings indicating more body dissatisfaction.

Bullying weight-related victimization. Two items from the revised Olweus Bully/Victim Questionnaire [57] adapted in the Italian context by Bacchini et al. [58] were used to assess physical appearance-based bullying perpetration ("I have teased someone about their physical appearance") and victimization ("I have been teased about my physical appearance") over the past six months. Responses were rated on a 5-point scale (1 = never; 2 = once/twice; 3 = 2/3 times a month; 4 = about once a week; 5 = several times a week), with higher scores indicating greater involvement in each role. The scale showed good internal consistency (ω = 0.848).

2.3. The Intervention Program

The intervention group participated in a six-session program, whereas the MIC group was exposed only to the program's introductory phase—a brief, 20 min session aimed at raising awareness about weight stigma.

The first phase of the intervention delivered targeted training to the class group, focusing on the concept of weight stigma, the various forms of peer violence—including bullying and body shaming—and the recognition of at-risk behaviors. The sessions explored the multiple manifestations of peer violence and discriminatory conduct that threaten adolescents' psychological, physical, moral, and social development, as well as their overall psychosocial well-being. Additionally, the training provided an overview of current national prevention, intervention, and victim protection frameworks, with particular attention to corresponding international and European systems, and guidelines for reporting harmful and discriminatory behaviors, especially those targeting children and adolescents.

The second phase, experiential in nature, engaged the class group in interactive workshops conducted through in-person sessions. These activities aimed to actively involve students in reflecting on the topics covered during the educational phase. Elements of the intervention drew on the principles of the validated Project Health protocol [42], which promotes healthy eating behaviors and active lifestyles. This approach encouraged participation in sports groups and the adoption of physically active routines. Concurrently, the program incorporated opportunities for peer bonding and group cohesion to cultivate an environment of inclusion and mutual acceptance.

The study team developed and implemented the intervention based on the biopsychosocial model of health [59], media literacy principles [60], and empirically supported strategies from the Project Health program [61]. The program comprised a total of six sessions, including two for pre- and post-assessment and four dedicated to the intervention (see below for a detailed description). Each lasting one hour per class, scheduled within school hours as either curricular or extracurricular activities, complemented by workshops and discussion groups. The number and length of sessions were adapted from existing evidence-based protocols (e.g., Project Health [42]) and balanced with the constraints of school timetables to optimise feasibility. Each session was facilitated by licensed psychologists. Regular supervisory meetings were conducted to ensure consistency and procedural fidelity across the sessions. A detailed preparatory training on the intervention framework was provided to facilitators, who also attended regular supervisory meetings to harmonise procedures and ensure consistency. Given the pilot and feasibility-oriented nature of the intervention, no formal manual or adherence checklists were employed. Fidelity was supported using detailed session scripts developed by the study team and delivered uniformly across classes.

Session 1: The initial session focused on introducing the concept of weight stigma, elucidating its definition, consequences, and the cyclical nature of stigmatization. Participants explored how stigma operates and its impact on individuals' psychological and social well-being.

Session 2: The second session addressed the behavioral components associated with weight stigma, including bullying and discrimination. Key contributing factors such as familial influences, social environments, and body image concerns were discussed in depth to understand their role in exacerbating stigmatising behaviors.

Session 3: The third session was dedicated to cognitive restructuring exercises aimed at challenging maladaptive thought patterns. Topics included cognitive avoidance strategies, practical advice to resist unrealistic beauty ideals, and engaging in cognitive-behavioral therapy (CBT) inspired strategies to challenge difficult situations.

Session 4: Four common myths related to eating habits, body image, obesity, and stigma were debunked, and students were encouraged to adopt healthy behaviors. A set of short educational video modules was developed and delivered as a final part of the intervention protocol. The modules were designed to provide participants with evidence-based insights into the complex, multifactorial nature of obesity, emphasizing its recognition as a chronic disease rather than a result of individual failure. Topics included the biopsychosocial model of health [62], the role of environmental and genetic factors in obesity, the impact of weight stigma on treatment outcomes, the influence of diet culture and the commercial diet industry, as well as the dismantling of common nutrition myths. This session was conducted from a weight-neutral health perspective, emphasizing that good health can be achieved across a range of body sizes when individuals engage in recommended nutrition and physical activity behaviors [12].

At the end of the sessions, in order to sum up all the information learned and be able to apply it in practice, participants were asked to write a supportive letter to a friend struggling with body image/weight-related concerns, thus fostering empathy and self-reflection.

At the end of the intervention phase and at the fulfilment of the second assessment, a celebratory event was planned for all the school divisions involved in the study. The celebration event was not part of the experimental content, and all the students from both the groups (intervention vs. MIC) participated. In collaboration with physical education teachers, recreational and team sports activities (e.g., tournaments) were organized to further support physical activity and social engagement. A t-shirt and booklet were distributed to all participants as the project's gadgets, including the slogans that some of the participating students had created to support the project's initiatives (e.g., "Stop body shaming!", "All the bodies are beautiful!", etc.). Particular attention was dedicated to the collection and distribution of the t-shirts (all sizes were available, making sure that everyone took their own independently and privately, see Figure 1 for a description of the program activities).

Students were also provided with an emergency support service for weight-based bullying, consisting of a no-cost helpline, available two days per week and operated in collaboration with a national child protection organization—Fondazione S.O.S Il Telefono Azzurro ETS—and a designated email address coordinated by the project team. This approach enabled continuous support throughout the intervention, ensuring that participants were not left unaided and were facilitated to disclose personal concerns or difficulties related to weight stigma and victimization (see Figure 2 for an overview of the proposed intervention).

Phase	Focus	Objectives
Educational Phase	Weight stigma, peer violence, and risk behaviors	Raising awareness, recognizing harmful behaviors,
Session 1	Definition and impact of weight stigma	Understanding how stigma affects well-being
Session 2	Bullying, discrimination, and contributing factors	Exploring the role of family, media and body image
Experiential Phase (Workshops & Physical Activities)	Group cohesion and inclusion, Healthy lifestyle habits	Enhancing psychosocial well-being, supporting physical activity and engagement
Session 3	Cognitive restructuring	Challenging negative thoughts and beauty ideals
Session 4	Myths on obesity and stigma	Debunk myths, promoting healthy attitudes and self-acceptance
Reflection Activity and Final Event	Supportive letter writing, Celebration with participating school classes	Fostering empathy and self-reflection, reinforcing belonging, celebrating achievements

Figure 1. The Structured Intervention Program.

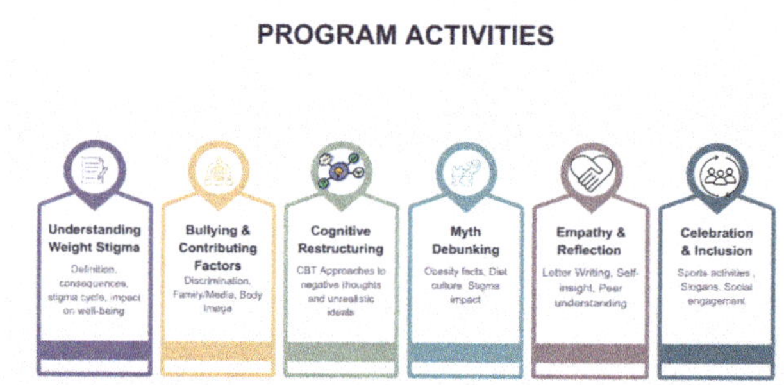

Figure 2. Overview of Proposed Intervention Activities.

2.4. The Psychoeducation Program for the MIC Group

School classes that were randomized in the MIC group received a 20 min psychoeducational session focused on the definition of weight stigma and its potential consequences on psychological well-being. This constituted a brief component of the initial intervention session administered to the intervention group and was implemented immediately before the pretest (T1). Students were encouraged to seek additional information, and they had continuous access to the free helpline throughout the duration of the program at the schools.

2.5. Data Analysis

Data were analysed using Jamovi v. 2.6.26 (https://www.jamovi.org/, accessed on 12 May 2025).

Firstly, socio-demographic descriptive statistics were calculated to characterise the sample, and bivariate correlations among the main study variables were calculated at baseline (T1).

Afterwards, four separate two-way mixed-design ANOVAs were performed to examine the effects of the intervention on WBIS, DEQ, ATOP, and CDRS scores, by considering Group (intervention vs. MIC) as the between-subjects factor and Moment of assessment (pre vs. post intervention) as the within-subjects factor. In the event of a significant Group × Moment interaction, analyses of simple main effects were planned to explore changes over time within each group separately.

To further explore the efficacy of the intervention and assess potential moderating effects, a series of exploratory ANCOVAs were conducted. For each outcome (WBIS, DEQ, ATOP, and CDRS), the post-treatment score (T2) was entered as the dependent variable, with Group (intervention vs. MIC) as the between-subjects factor and the corresponding baseline score (T1) as a covariate. In addition, BMI and sex were included as covariates, and their interactions with Group were tested to examine potential moderating effects. This approach allowed us to assess whether the intervention effect was consistent across BMI and sex, and to control for potential baseline differences in the outcomes.

Effect sizes were expressed using partial eta squared (ηp^2), with values of 0.01, 0.06, and 0.14 denoting small, moderate, and large effects, respectively [63]. The α level was set at 0.05 to determine statistical significance.

3. Results

3.1. Socio-Demographic Characteristics

A total of 539 adolescents completed the pre-intervention assessment and were enrolled in the study. Of these, 38.8% (n = 209) were in the MIC condition and 61.2% (n = 330) in the intervention condition. Participants had a mean age of 15.91 years (SD = 1.38; range: 13–21) and were sex-balanced (51.5% males and 48.5% females). Of the total sample, 21.89% (comparison group: n = 53; experimental group: n = 65) reported having teased someone about their weight or physical appearance at least once in the past six months, while 22% (comparison group: n = 44; experimental group: n = 75) reported having personally experienced teasing for the same reasons during the same period. Notably, 25.2% of the sample reported having experienced body shaming by family members, while 26.8% reported experiencing body shaming by peers.

Moreover, a total of 73 adolescents self-identified in a condition of overweight or obesity (Mean BMI = 26.49 kg/m^2, SD = 4.04) and completed the WBIS. This subgroup had a mean age of 15.81 years (SD = 1.31; range: 14–19) and was predominantly female (65.8%). Critically, 53.4% of these participants reported having experienced body shaming by family members, while 57.5% reported experiencing body shaming by peers. Of this subgroup, 31.5% (n = 23) were assigned to the MIC condition and 68.5% (n = 50) to the intervention condition.

Detailed descriptive statistics of the investigated variables are summarised in Table 1.

3.2. Bivariate Correlations

Pearson's correlation coefficients representing bivariate correlations among variables measured at baseline (T1) are summarized in Table 2. More specifically, DEQ scores were strongly correlated with WBIS (r = 0.700, $p < 0.001$) and CDRS (r = −0.610, $p < 0.001$) scores. Moreover, WBIS scores were significantly and moderately correlated with CDRS scores (r = −0.373, $p < 0.001$).

Table 1. Means (M) and standard errors (SE) by experimental condition and time.

Variable	Time 1		Time 2	
	MIC Group (n = 209) M (SE)	Intervention Group (n = 330) M (SE)	MIC Group (n = 209) M (SE)	Intervention Group (n = 330) M (SE)
WBIS	45.043 (2.988)	53.140 (2.027)	41.870 (3.271)	49.020 (2.218)
DEQ	27.751 (1.702)	30.110 (1.359)	28.276 (1.703)	30.407 (1.360)
ATOP	63.236 (1.006)	63.036 (0.796)	63.115 (1.042)	63.236 (0.824)
CDRS	−0.822 (0.132)	−0.875 (0.106)	−0.908 (0.127)	−0.787 (0.102)

Notes. DEQ: Disordered Eating Questionnaire; WBIS: Weight Bias Internalization Scale; ATOP: Attitudes Toward Obese Persons; CDRS: Contour Drawing Rating Scale.

Table 2. Bivariate correlations among variables measured at baseline (T1, n = 539).

	DEQ	WBIS	ATOP	SRS
DEQ	1	-	-	-
WBIS	0.700 *	1	-	-
ATOP	0.016	0.021	1	
CDRS	−0.610 *	−0.373 *	−0.015	1

Note. DEQ: Disordered Eating Questionnaire; WBIS: Weight Bias Internalization Scale; ATOP: Attitudes Toward Obese Persons; CDRS: Contour Drawing Rating Scale. * $p < 0.001$.

3.3. *Two-Way Mixed-Design ANOVAs*

Two-way mixed-design ANOVAs were implemented with Group (intervention vs. MIC) as the between-subjects factor and Moment (pre vs. post intervention) as the within-subjects factor. Four distinct analyses were conducted, considering WBIS, DEQ, ATOP, and CDRS scores as dependent variables, respectively. Detailed results are reported in Supplementary Table S1.

3.3.1. Attitudes Toward Obese Persons Scores

ANOVA revealed no significant Group × Moment interaction, F(1, 494) = 0.076, p = 0.783, ηp^2 = 0.000, indicating that the trajectory of ATOP scores from pre- to post-intervention did not differ between groups. Similarly, no significant main effects emerged for Moment (F(1, 494) = 0.005, p = 0.945, ηp^2 = 0.000) or Group (F(1, 494) = 0.001, p = 0.973, ηp^2 = 0.000), suggesting that neither time nor condition influenced participants' attitudes toward obese persons.

3.3.2. Weight Bias Internalization Scale Scores

Findings revealed no significant Group × Moment interaction, F(1, 71) = 0.126, p = 0.723, ηp^2 = 0.002, indicating that changes in WBIS scores over time did not differ significantly between the two experimental conditions. Nonetheless, significant main effects of Group (F(1, 71) = 4.627, p = 0.035, ηp^2 = 0.061) and Moment (F(1, 71) = 7.518, p = 0.008, ηp^2 = 0.096) were observed, suggesting that, on average, participants in the intervention and MIC groups differed in their levels of internalized weight bias, and that WBIS scores changed significantly from pre- to post-intervention, regardless of group membership.

To further explore the significant main effects, descriptive statistics were examined. On average, participants in the intervention group reported higher WBIS scores (M = 51.08, SE = 1.98) compared to those in the MIC group (M = 43.45, SE = 2.93), across both time points. Furthermore, WBIS scores decreased from pre (M = 49.09, SE = 1.80) to post (M = 45.44, SE = 1.97) intervention, suggesting that participants experienced a reduction in internalized weight bias over time, regardless of whether they were in the intervention or MIC condition. These patterns are graphically illustrated in Figure 3.

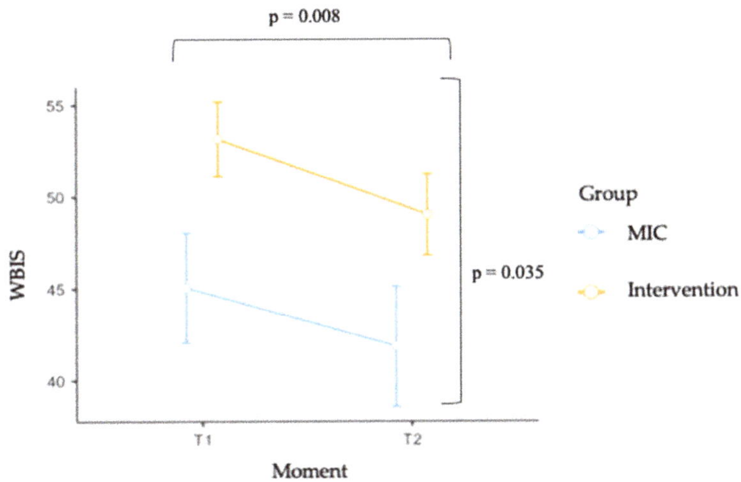

Figure 3. WBIS scores by group (MIC vs. Intervention) across time (Moment). Error bars represent standard errors of the mean (SE). Square brackets indicate the main effects of Group and Time.

3.3.3. Disordered Eating Questionnaire Scores

Findings revealed no significant Group × Moment interaction, F(1, 473) = 0.025, p = 0.875, ηp^2 = 0.000, indicating that changes in DEQ scores over time did not differ between the two experimental conditions. Moreover, no significant main effects of Moment (F(1, 473) = 0.324, p = 0.569, ηp^2 = 0.001) or Group (F(1, 473) = 1.193, p = 0.275, ηp^2 = 0.003) were found.

3.3.4. Contour Drawing Rating Scale Scores

Interestingly, the analysis conducted on CDRS scores revealed a marginally significant Group × Moment interaction, F(1, 470) = 3.152, p = 0.076, ηp^2 = 0.007, indicating that changes in body size dissatisfaction over time marginally differ between the MIC and intervention groups. Furthermore, no significant main effects of Moment (F(1, 470) = 0.001, p = 0.995, ηp^2 = 0.000) or Group (F(1, 470) = 0.046, p = 0.831, ηp^2 = 0.000) were found. The interaction effect is illustrated in Figure 4. The effect in the intervention group did not attain statistical significance (p > 0.05), yet a consistent direction of change was apparent, with the discrepancy between actual and desired body size decreasing from pre-intervention (M = −0.875, SE = 0.106) to post-intervention (M = −0.787, SE = 0.102).

3.4. Exploratory ANCOVAs

To further explore intervention efficacy while accounting for baseline differences, a series of ANCOVAs were conducted. For each outcome measure (WBIS, DEQ, ATOP, and CDRS), the post-treatment score (T2) was entered as the dependent variable, with Group (intervention vs. MIC) as the between-subjects factor. Baseline scores (T1), BMI, and sex were included as covariates. Interaction terms (Group × BMI and Group × Sex) were also included to examine their potential moderating effects.

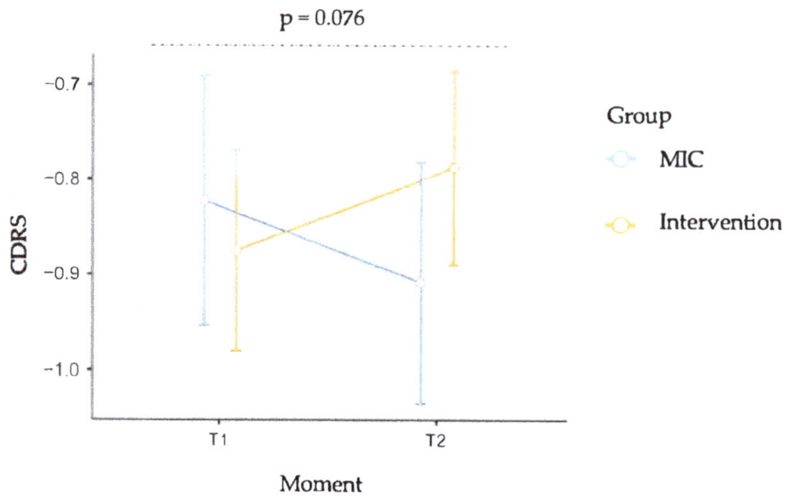

Figure 4. Contour Drawing Rating Scale (CDRS) scores by group (MIC vs. Intervention) across time (Moment). Error bars represent standard errors of the mean (SE). The dashed line indicates the Group × Moment interaction.

3.4.1. Attitudes Toward Obese Persons Scores

ANCOVA revealed no significant Group × Sex interaction for ATOP scores at T2, F(1, 473) = 3.309, $p = 0.070$, $\eta p^2 = 0.007$, suggesting that the effect of the intervention did not significantly differ by sex. Similarly, the Group × BMI interaction was not significant (F(1, 473) = 1.170, $p = 0.280$, $\eta p^2 = 0.002$). The main effect of Group also remained non-significant, F(1, 473) = 0.148, $p = 0.701$, $\eta p^2 = 0.000$, indicating that, after controlling for baseline scores, BMI, and sex, the intervention did not yield a significant change in ATOP scores compared to the MIC group.

3.4.2. Weight Bias Internalisation Scale Scores

ANCOVA showed no significant Group × Sex interaction for WBIS scores at T2, F(1, 63) = 0.419, $p = 0.520$, $\eta p^2 = 0.007$, indicating that the intervention effect did not differ significantly between sexes. The interaction between Group and BMI was also non-significant (F(1, 63) = 0.079, $p = 0.779$, $\eta p^2 = 0.001$). Similarly, the main effect of Group was meaningless, F(1, 63) = 0.207, $p = 0.651$, $\eta p^2 = 0.003$, suggesting that, after accounting for baseline scores, BMI, and sex, the intervention had no significant impact on post-intervention WBIS scores compared to the MIC condition.

3.4.3. Disordered Eating Questionnaire Scores

ANCOVA did not reveal a significant Group × Sex interaction on DEQ scores at T2, F(1, 452) = 0.828, $p = 0.363$, $\eta p^2 = 0.002$, suggesting that the intervention effect was not moderated by sex. Likewise, no significant interaction emerged between Group and BMI (F(1, 452) = 0.205, $p = 0.651$, $\eta p^2 = 0.000$). The main effect of Group was also negligible, F(1, 452) = 0.018, $p = 0.893$, $\eta p^2 = 0.000$, indicating that, when controlling for baseline scores, BMI, and sex, the intervention exerted no influence on post-intervention DEQ scores compared to the MIC condition.

3.4.4. Contour Drawing Rating Scale Scores

ANCOVA did not reveal a significant Group × Sex interaction on CDRS scores at T2, F(1, 450) = 0.001, $p = 0.996$, $\eta p^2 = 0.000$, suggesting that the effect of the intervention did not differ by sex. Similarly, the Group × BMI interaction was not significant ($F(1, 450) = 0.249$, $p = 0.618$, $\eta p^2 = 0.001$). Importantly, the main effect of Group approached statistical significance, $F(1, 450) = 3.831$, $p = 0.051$, $\eta p^2 = 0.008$. Estimated marginal means showed that, after adjusting for baseline scores, BMI, and sex, participants in the intervention group reported a smaller discrepancy between actual and desired body size at post-intervention (M = −0.727, SE = 0.058) compared to those in the MIC group (M = −0.905, SE = 0.070).

4. Discussion

The findings of this study offer valuable insights into both the potential and limitations of a brief intervention aimed at reducing weight stigma and related aspects in high school students. Firstly, results highlight that 13.54% (*n* = 72) of the total sample self-identified as being in a condition of OB/OW. Importantly, more than half of these participants reported having experienced body shaming by family members and peers. This evidence aligns well with previous research highlighting the potential role of self-weight perception in influencing adolescents' susceptibility to weight-based victimisations and suggesting that intervention efforts should address not only actual body weight, but also self-perceptions and concerns related to body image. Moreover, this finding confirmed previous evidence on the prevalence of weight bias among adolescents [64], especially in educational environments [65,66], and established school context as a fertile ground for the development of weight-based discrimination. Additionally, descriptive results suggest that more than 20% of the total sample reported having perpetrated or experienced appearance-based bullying. These findings confirm the pervasive nature of appearance and weight-based bullying–either acted as perpetrator or victim-among adolescents (e.g., [67]) and highlight the need for interventions designed to address both the victimization and the perpetration of these forms of bullying. Previous studies reported that between 11% and 54% of Italian school students experienced teasing related to physical appearance, with higher rates observed among students with more severe obesity [67]. Future research should investigate whether the effects of the intervention vary across students with different weight profiles. While appearance-related bullying is common, other forms of bullying (e.g., based on race, ethnicity, religion, sexual orientation) tend to have lower prevalence rates, generally ranging between 5% and 15% in adolescent populations [68,69]. This comparison highlights that, although all forms of bullying are problematic, appearance and weight remain among the most frequent triggers for peer victimisation.

This is true especially considering the developmental period of adolescence, characterised by profound physical, psychological, and social changes [70]. During this transitional period, a series of developmental processes occur, including increases in self-consciousness and awareness of external and sociocultural expectations [71]. Given the heightened relevance of peer approval and social conformity during adolescence, this population can be distinctively susceptible to questioning their self-worth and thus may develop weight bias in response to external sociocultural pressures [30,44]. Indeed, adolescents are particularly susceptible to weight stigma and its harmful effects, such as teasing, bullying, social isolation, physical health issues, and emotional distress [30]. One of the main contributors to weight stigma during adolescence is social media, which is widely used by teens around the world. These platforms often expose users to weight-related messages that tend to marginalise individuals with higher body weights, fostering the internalisation of weight bias [72]. For adolescents, experiencing weight stigma has been linked to higher rates of depression, anxiety, suicidal thoughts, disordered eating behaviors, and substance use [32]. Future works are needed to investigate potential risk factors and their reciprocal relationship influencing the structuration of weight-based stereotypes in youth, as well as potential health-related consequences.

Emerging evidence indicates that the phenomenon is also quite widespread in Italy, and experiences of weight stigmatisation, bullying, and body shaming are frequently reported in association with poorer health outcomes [47,48,67]. Our school-based intervention program was designed to reduce weight stigma and weight-related teasing through the face-to-face delivery of a four-session training, conducted by researchers who were all licensed psychologists or psychotherapists with expertise in this area.

The evaluation of our intervention program was primarily focused on the comparison between pre- and post-intervention scores of internalized weight stigma, alongside associated aspects including body dissatisfaction, eating symptomatology, and weight-based attitudes. Overall, no statistically significant effects of the intervention emerged for the majority of aspects considered, and the results should be interpreted with caution. The intervention yielded small effect sizes across the analyses, suggesting that the proposed program produced limited and heterogeneous changes rather than robust, sustained reductions in weight stigma and related factors.

More specifically, no significant intervention effects were found for explicit attitudes towards obesity (i.e., ATOP). Previous evidence emphasises a small to medium effect of existing interventions on explicit weight-biased attitudes and beliefs [73], implying that short-term interventions may be insufficient to alter these aspects in adolescents. However, in the subgroup of participants perceiving themselves as being in an overweight condition, a significant effect of Time occurred in predicting a reduction in internalized weight stigma (i.e., WBIS). The absence of a Group × Moment effect indicated similar changes in both groups—MIC and experimental—over time. Multiple explanations could be mentioned for this result. First, the low-intensity psychoeducation provided to the MIC group at the pre- and post-assessment may have been beneficial, suggesting that even a minimal education exposure could exert effects in reducing weight bias internalisation, similarly to a complete intervention protocol [29]. Second, the educators of students in both conditions were recommended to pay attention to weight stigma topics, in line with the intervention protocol, as suggested elsewhere [66]. This may have fostered the reduction in weight stigma across all the students involved. Future studies are needed to assess the impacts of teachers' weight-related attitudes on students' own perceptions and attitudes toward weight. Moreover, the main significant contribution of the Group indicated that internalized weight bias was generally higher in the intervention group, irrespective of the time point, suggesting that participants in the experimental condition may have exhibited a higher baseline level of weight bias, a trajectory which was maintained in post-intervention scores. This could reflect a "baseline balance effect" [74] and should be addressed as one limitation of this study. Furthermore, the sole contribution of time in predicting weight bias reduction in both groups may agree with the "Hawthorne effect", a non-specific treatment effect where the awareness of being assessed may alter the participants' responses and behaviors [75].

Results on body dissatisfaction revealed a marginally significant Group × Moment interaction effect, indicating that changes in this aspect over time marginally differ between the MIC and intervention groups. The effect did not reach statistical significance, and therefore must be interpreted cautiously, although a small reduction in the discrepancy between actual and desired body size was observed from pre- to post-intervention in the experimental group. This modest change aligns with previous observations [29] but should be interpreted with caution. While this finding, albeit limited, could encourage the potential value of integrating weight stigma-focused content in existing intervention protocols designed to improve body image concerns, alternative explanations of these effects—such as short intervention duration or sample characteristics—cannot be excluded. Considering the role of overvaluation of shape and weight as a core psychopathological determinant of eating disorders [76,77], further research is needed to determine whether more intensive programs could yield meaningful improvements.

No significant intervention effects were found for eating symptomatology (i.e., DEQ). Previous evidence of substantial improvement in eating disturbances after a weight-stigma focused intervention suggests heterogeneous results [78]. One potential explanation of this negligible effect could be the brief and limited duration of the present program, together with the fact that most weight bias intervention protocols mainly focus on attitudes rather than emotional and behavioral mechanisms underpinning eating symptomatology [78]. Future experimental studies should address this limitation by integrating these aspects into the intervention protocols.

This study clearly presented some limitations. First, the mere employment of self-reported measures may have constrained the robustness of the assessment, potentially enhancing the desirability bias. Future studies should consider integrating objective measures, such as the implicit association test (IAT) for weight bias [79], an experimental tool that captures automatic negative weight-based attitudes. Second, the inclusion of Italian secondary school students may have influenced the generalizability of findings due to cultural or educational contexts. For example, the Italian school system is characterised by teacher-led classroom activities, which may have supported the delivery of psychoeducational modules. Third, the heterogeneity of participant baseline weight bias levels limited the robustness of conclusions on intervention efficacy as well as their internal validity. Further research could employ blocking or stratification based on key baseline characteristics (e.g., weight stigma levels) to improve group balance. Fourth, randomisation was performed at the class rather than the individual level; the number of participants allocated to each condition was determined by pre-existing differences in class enrollment, leading to uneven group sizes. Moreover, a contamination of information may have taken place from the intervention group to the MIC group, due to potential informal communication between students across classes, which also may have diluted between-group differences. Future studies should address this issue by adopting a cluster-randomised design at the school level, thereby minimising the likelihood of cross-group information exchange. Sixth, since the control group received minimal psychoeducation on weight stigma, a true control condition was not established. Future studies could include an active control group with content unrelated to weight stigma or a waitlist control group to further strengthen internal validity. Furthermore, the absence of systematic attendance records limited our ability to examine potential dose–response effects of the intervention, which may have influenced the magnitude of the observed outcomes. Future studies should assess this aspect by recording adherence to the intervention sessions for each participant. Finally, the lack of a follow-up assessment hinders the evaluation of the intervention in the long term and the maintenance of the short-term effects over time. Moreover, this prevented the observation of potential score regression at follow-up in the MIC group. Future studies should address this point by including more occasions of post-intervention assessment. Finally, future studies should enhance intervention fidelity by employing standardised manuals and formal adherence checklists in addition to facilitator training and supervision.

Notwithstanding these limitations, the present study possesses the distinct strength of providing a unique and significant contribution to the scientific literature on weight stigma interventions, offering valuable recommendations for the development of future prevention and experimental protocols for a reduction in weight stigma and related aspects in educational settings. The principal strength of this intervention resides in its grounding on the international recommendation of the World Obesity Federation to reduce weight stigma (for a detailed description, see Nutter et al. [80]). In line with the position paper, our program tries to change the common social narratives, language, and images used to discuss body weight, obesity, and appearance and specifically: (1) uses non-stigmatizing language and imagery; (2) uses person-first language (describing what a person "has" rather than asserting what a person "is"); (3) engages in weight-neutral health promotion, especially in the sections targeting physical activity and nutritional aspect (e.g., the set of short educational video modules which focuses on health promotion strategies defocusing on weight, weight loss, and "healthy weight" (based on BMI); and (4) responds to the need of raising awareness of weight stigma and increasing the global evidence base [80]. It is our aspiration that this pilot intervention program may serve to develop a more effective protocol to reduce weight stigma and related aspects, ultimately improving the psychological and social well-being of adolescents.

Supplementary Materials: The following supporting information can be downloaded at: https://www.mdpi.com/article/10.3390/children12091208/s1, Table S1: Detailed results from two-way mixed ANOVAs. Note: * $p < 0.05$; † $p < 0.10$. Abbreviations: DEQ: Disordered Eating Questionnaire; WBIS: Weight Bias Internalization Scale; ATOP: Attitudes Toward Obese Persons; CDRS: Contour Drawing Rating Scale; SE: standard error.

Author Contributions: Conceptualization, M.V., S.C., A.Z. (Anna Zegretti) and C.L.; methodology, M.V., S.C., A.Z. (Anna Zegretti), A.Z. (Andrea Zagaria) and C.L.; validation, M.V., S.C. and C.L.; formal analysis, M.V., S.C. and A.Z. (Andrea Zagaria); investigation, M.V., S.C. and A.Z. (Anna Zegretti); resources, S.C., A.Z. (Anna Zegretti), A.Z. (Andrea Zagaria), E.M. and C.L.; data curation, M.V., A.Z. (Andrea Zagaria) and S.C.; writing—original draft preparation, M.V.; writing—review and editing, S.C., A.Z. (Anna Zegretti) and C.L.; visualization, S.C., A.Z. (Anna Zegretti), A.Z. (Andrea Zagaria), E.M. and C.L. supervision, project administration, C.L.; funding acquisition, S.C., A.Z. (Anna Zegretti) and C.L. All authors have read and agreed to the published version of the manuscript.

Funding: This research was supported by a research grant from the Italian Department for Family Policies under the Presidency of the Council of Ministers (Grant CUP B85F21003150001; Project title: Prevention and Counteraction of Weight-Based Bullying: Sport and Physical Exercise for a Multilevel Intervention; Approval date: 15 June 2021).

Institutional Review Board Statement: The study was conducted in accordance with the Declaration of Helsinki and approved by the Institutional Review Board of the Department of Psychology at Sapienza University of Rome (date: 27 May 2022; prot. Number 0001069).

Informed Consent Statement: Informed consent was obtained from all subjects involved in the study.

Data Availability Statement: The data presented in this study are available on request from the corresponding author due to the privacy and confidentiality of participants.

Acknowledgments: This manuscript was prepared with the support of AI-assisted tools for bibliography management.

Conflicts of Interest: The authors declare no conflicts of interest.

References

1. Carl, J.; Thedinga, H.K.; Zipfel, S.; Thiel, A. Stimulating weight stigma in future experimental designs on physical activity—Development and pilot validation of a video instrument. *Obes. Facts* **2018**, *11*, 206–220. [CrossRef] [PubMed]
2. Brochu, P.M.; Esses, V.M. What's in a name? The effects of the labels "fat" versus "overweight" on weight bias. *J. Appl. Soc. Psychol.* **2011**, *41*, 1981–2008. [CrossRef]
3. Puhl, R.M.; Moss-Racusin, C.A.; Schwartz, M.B.; Brownell, K.D. Weight stigmatization and bias reduction: Perspectives of overweight and obese adults. *Health Educ. Res.* **2007**, *23*, 347–358. [CrossRef]

4. Kopp, W. How Western diet and lifestyle drive the pandemic of obesity and civilization diseases. *Diabetes Metab. Syndr. Obes.* **2019**, *12*, 2221–2236. [CrossRef]
5. Chooi, Y.C.; Ding, C.; Magkos, F. The epidemiology of obesity. *Metabolism* **2019**, *92*, 6–10. [CrossRef]
6. Whitlock, G.; Lewington, S.; Sherliker, P.; Clarke, R.; Emberson, J.; Halsey, J.; Qizilbash, N.; Collins, R.; Peto, R. Body-mass index and cause-specific mortality in 900,000 adults: Collaborative analyses of 57 prospective studies. *Lancet* **2009**, *373*, 1083–1096. [PubMed]
7. Lauro, R.; Sbraccia, P.; Lenzi, A. 1st Italian Obesity Barometer Report. *Obes. Monit.* **2019**, *1*, 1–88.
8. Istituto Superiore di Sanità. Okkio Alla Salute. Available online: http://www.epicentro.iss.it/okkioallasalute (accessed on 31 August 2025).
9. Istituto Nazionale di Statistica; Bologna, E. Fattori di Rischio per la Salute: Fumo, Obesità, Alcol e Sedentarietà. Statistiche Report 2016. Available online: http://www.istat.it/it/archivio/189498 (accessed on 31 August 2025).
10. De Carli, L.; Spada, E.; Milani, S.; Ganzit, G.P.; Ghizzoni, L.; Raia, M.; Coscia, A.; Bertino, E.; Bo, S. Prevalence of obesity in Italian adolescents: Does the use of different growth charts make the difference? *World J. Pediatr.* **2018**, *14*, 168–175. [CrossRef]
11. Puhl, R.M.; Himmelstein, M.S.; Quinn, D.M. Internalizing weight stigma: Prevalence and sociodemographic considerations in US adults. *Obesity* **2017**, *26*, 167–175. [CrossRef]
12. Smith, G.I.; Mittendorfer, B.; Klein, S. Metabolically healthy obesity: Facts and fantasies. *J. Clin. Investig.* **2019**, *129*, 3978–3989. [CrossRef]
13. World Health Organization; Mathers, C. Global Strategy for Women's, Children's and Adolescents' Health (2016–2030). WHO 2015. Available online: https://apps.who.int/gb/ebwha/pdf_files/WHA72/A72_30-en.pdf (accessed on 31 August 2025).
14. Voelker, D.K.; Reel, J.J.; Greenleaf, C. Weight status and body image perceptions in adolescents: Current perspectives. *Adolesc. Health Med. Ther.* **2015**, *6*, 149–158.
15. Keijsers, L.; Voelkle, M.C.; Maciejewski, D.; Branje, S.; Koot, H.; Hiemstra, M.; Meeus, W. What drives developmental change in adolescent disclosure and maternal knowledge? Heterogeneity in within-family processes. *Dev. Psychol.* **2016**, *52*, 2057–2070. [CrossRef]
16. Harter, S. *Self-Perception Profile for Adolescents: Manual and Questionnaires*; University of Denver, Department of Psychology: Denver, CO, USA, 2012; pp. 31–45.
17. Altabe, M.; Thompson, J.K. Body image: A cognitive self-schema construct? *Cognit. Ther. Res.* **1996**, *20*, 171–193. [CrossRef]
18. Clay, D.; Vignoles, V.L.; Dittmar, H. Body image and self-esteem among adolescent girls: Testing the influence of sociocultural factors. *J. Res. Adolesc.* **2005**, *15*, 451–477. [CrossRef]
19. Puhl, R.M.; Wall, M.M.; Chen, C.; Austin, S.B.; Eisenberg, M.E.; Neumark-Sztainer, D. Experiences of weight teasing in adolescence and weight-related outcomes in adulthood: A 15-year longitudinal study. *Prev. Med.* **2017**, *100*, 173–179. [CrossRef]
20. Himmelstein, M.S.; Puhl, R.M. Weight-based victimization from friends and family: Implications for how adolescents cope with weight stigma. *Pediatr. Obes.* **2019**, *14*, e12453. [CrossRef]
21. Puhl, R.M.; Luedicke, J. Weight-based victimization among adolescents in the school setting: Emotional reactions and coping behaviors. *J. Youth Adolesc.* **2011**, *41*, 27–40. [CrossRef]
22. Markey, C.N. Invited commentary: Why body image is important to adolescent development. *J. Youth Adolesc.* **2010**, *39*, 1387–1391. [CrossRef] [PubMed]
23. Holland, G.; Tiggemann, M. A systematic review of the impact of the use of social networking sites on body image and disordered eating outcomes. *Body Image* **2016**, *17*, 100–110. [CrossRef]
24. Gmeiner, M.S.; Warschburger, P. Interrelation between weight and weight stigma in youth: Is there evidence for an obesogenic vicious cycle. *Eur. Child Adolesc. Psychiatry* **2023**, *32*, 697–704. [CrossRef] [PubMed]
25. Ratcliffe, D.; Ellison, N. Obesity and internalized weight stigma: A formulation model for an emerging psychological problem. *Behav. Cogn. Psychother.* **2013**, *43*, 239–252. [CrossRef]
26. Durso, L.E.; Latner, J.D. Understanding self-directed stigma: Development of the weight bias internalization scale. *Obesity* **2008**, *16*, S80–S86. [CrossRef]
27. Golaszewski, N.M.; Pasch, K.E.; Fernandez, A.; Poulos, N.S.; Batanova, M.; Loukas, A. Perceived weight discrimination and school connectedness among youth: Does teacher support play a protective role? *J. Sch. Health* **2018**, *88*, 754–761. [CrossRef]
28. Wee, C.C.; Davis, R.B.; Huskey, K.W.; Jones, D.B.; Hamel, M.B. Quality of life among obese patients seeking weight loss surgery: The importance of obesity-related social stigma and functional status. *J. Gen. Intern. Med.* **2012**, *28*, 231–238. [CrossRef]
29. Puhl, R.; Suh, Y. Health consequences of weight stigma: Implications for obesity prevention and treatment. *Curr. Obes. Rep.* **2015**, *4*, 182–190. [CrossRef]
30. Pont, S.J.; Puhl, R.; Cook, S.R.; Slusser, W. Stigma experienced by children and adolescents with obesity. *Pediatrics* **2017**, *140*, e20173034. [CrossRef]
31. Tomiyama, A.J. Weight stigma is stressful: A review of evidence for the cyclic obesity/weight-based stigma model. *Appetite* **2014**, *82*, 8–15. [CrossRef] [PubMed]

32. Puhl, R.M.; Himmelstein, M.S.; Pearl, R.L. Weight stigma as a psychosocial contributor to obesity. *Am. Psychol.* **2020**, *75*, 274–289. [CrossRef]
33. Brown, A.; Flint, S.W.; Batterham, R.L. Pervasiveness, impact and implications of weight stigma. *EClinicalMedicine* **2022**, *47*, 101408. [CrossRef] [PubMed]
34. Roberts, K.J.; Polfuss, M.L.; Marston, E.C.; Davis, R.L. Experiences of weight stigma in adolescents with severe obesity and their families. *J. Adv. Nurs.* **2021**, *77*, 4184–4194. [CrossRef] [PubMed]
35. Diedrichs, P.C.; Barlow, F.K. How to lose weight bias fast! Evaluating a brief anti-weight bias intervention. *Br. J. Health Psychol.* **2011**, *16*, 846–861. [CrossRef]
36. Puhl, R.M.; Schwartz, M.B.; Brownell, K.D. Impact of perceived consensus on stereotypes about obese people: A new approach for reducing bias. *Health Psychol.* **2005**, *24*, 517–525. [CrossRef]
37. Puhl, R.M.; Heuer, C.A. The stigma of obesity: A review and update. *Obesity* **2009**, *17*, 941–964. [CrossRef] [PubMed]
38. Batson, C.D.; Ahmad, N.Y. Using empathy to improve intergroup attitudes and relations. *Soc. Issues Policy Rev.* **2009**, *3*, 141–177. [CrossRef]
39. Griffiths, L.J.; Page, A.S. The impact of weight-related victimization on peer relationships: The female adolescent perspective. *Obesity* **2008**, *16*, S39–S45. [CrossRef] [PubMed]
40. Miyairi, M.; Reel, J.J.; Próspero, M.; Okang, E.N. Addressing size stereotypes: A weight bias and weight-related teasing intervention among adolescents. *J. Health Educ. Teach.* **2015**, *6*, 33–42.
41. Bell, S.; Morgan, M. Children's attitudes and behavioral intentions toward a peer presented as obese: Does a medical explanation make a difference? *J. Pediatr. Psychol.* **2000**, *25*, 137–145. [CrossRef]
42. Stice, E.; Rohde, P.; Shaw, H.; Gau, J.M. An experimental therapeutics test of whether adding dissonance-induction activities improves the effectiveness of a selective obesity and eating disorder prevention program. *Int. J. Obes.* **2018**, *42*, 462–468. [CrossRef]
43. Nutter, S.; Alberga, A.S.; MacInnis, C.; Brun, I.; Lefebvre, D.; Hayden, K.A.; Russell-Mayhew, S. Framing obesity a disease: Indirect effects of affect and controllability beliefs on weight bias. *Int. J. Obes.* **2018**, *42*, 1804–1811. [CrossRef]
44. Puhl, R.M.; Lessard, L.M. Weight stigma in youth: Prevalence, consequences, and considerations for clinical practice. *Curr. Obes. Rep.* **2020**, *9*, 402–411. [CrossRef]
45. Pearl, R.L.; Hopkins, C.H.; Berkowitz, R.I.; Wadden, T.A. Group cognitive-behavioral treatment for internalized weight stigma: A pilot study. *Eat. Weight Disord.* **2018**, *23*, 357–362. [CrossRef]
46. Puhl, R.M.; Luedicke, J.; Heuer, C. Weight-based victimization toward overweight adolescents: Observations and reactions of peers. *J. Sch. Health* **2011**, *81*, 696–703. [CrossRef]
47. Cerolini, S.; Vacca, M.; Zegretti, A.; Zagaria, A.; Lombardo, C. Body shaming and internalized weight bias as potential precursors of eating disorders in adolescents. *Front. Psychol.* **2024**, *15*, 1356647. [CrossRef]
48. Vacca, M.; Cerolini, S.; Zegretti, A.; Zagaria, A.; Lombardo, C. Bullying victimization and adolescent depression, anxiety and stress: The mediation of cognitive emotion regulation. *Children* **2023**, *10*, 1897. [CrossRef]
49. Pearl, R.L.; Puhl, R.M. Measuring internalized weight attitudes across body weight categories: Validation of the modified weight bias internalization scale. *Body Image* **2014**, *11*, 89–92. [CrossRef]
50. Innamorati, M.; Imperatori, C.; Lamis, D.A.; Contardi, A.; Castelnuovo, G.; Tamburello, S.; Fabbricatore, M. Weight Bias Internalization Scale discriminates obese and overweight patients with different severity levels of depression: The Italian version of the WBIS. *Curr. Psychol.* **2017**, *36*, 242–251. [CrossRef]
51. Allison, D.B.; Basile, V.C.; Yuker, H.E. The measurement of attitudes toward and beliefs about obese persons. *Int. J. Eat. Disord.* **1991**, *10*, 599–607. [CrossRef]
52. Zagaria, A.; Mocini, E.; Cerolini, S.; Donini, L.M.; Castelnuovo, G.; Manzoni, G.M.; Lombardo, C. A validation study of the Italian version of the attitudes toward obese persons (I-ATOP) questionnaire. *Obes. Res. Clin. Pract.* **2022**, *16*, 262–268. [CrossRef] [PubMed]
53. Tsai, M.C.; Strong, C.; Latner, J.D.; Lin, Y.C.; Pakpour, A.H.; Lin, C.Y.; Wang, S.M. Attitudes toward and beliefs about obese persons across Hong Kong and Taiwan: Wording effects and measurement invariance. *Health Qual. Life Outcomes* **2019**, *17*, 134. [CrossRef] [PubMed]
54. Lombardo, C.; Russo, P.M.; Lucidi, F.; Iani, L.; Violani, C. Internal consistency, convergent validity and reliability of a brief Questionnaire on Disordered Eating (DEQ). *Eat. Weight Disord.* **2004**, *9*, 91–98. [CrossRef] [PubMed]
55. Lombardo, C.; Cuzzolaro, M.; Vetrone, G.; Mallia, L.; Violani, C. Concurrent validity of the Disordered Eating Questionnaire (DEQ) with the Eating Disorder Examination (EDE) clinical interview in clinical and non-clinical samples. *Eat. Weight Disord.* **2011**, *16*, e188–e198. [CrossRef] [PubMed]
56. Thompson, M.A.; Gray, J.J. Development and validation of a new body-image assessment scale. *J. Pers. Assess.* **1995**, *64*, 258–269. [CrossRef]
57. Solberg, M.E.; Olweus, D. Prevalence estimation of school bullying with the Olweus Bully/Victim Questionnaire. *Aggress. Behav.* **2003**, *29*, 239–268. [CrossRef]

58. Bacchini, D.; Esposito, G.; Affuso, G. Social experience and school bullying. *J. Community Appl. Soc. Psychol.* **2009**, *19*, 17–32. [CrossRef]
59. Engel, G.L. The need for a new medical model: A challenge for biomedicine. *Science* **1977**, *196*, 129–136. [CrossRef] [PubMed]
60. Potter, W.J. *Media Literacy*; Sage Publications: Thousand Oaks, CA, USA, 2018.
61. Stice, E.; Shaw, H.; Burton, E.; Wade, E. Dissonance and healthy weight eating disorder prevention programs: A randomized efficacy trial. *J. Consult. Clin. Psychol.* **2006**, *74*, 263–275. [CrossRef]
62. Lehman, B.J.; David, D.M.; Gruber, J.A. Rethinking the biopsychosocial model of health: Understanding health as a dynamic system. *Soc. Personal. Psychol. Compass* **2017**, *11*, e12328. [CrossRef]
63. Cohen, J. *Statistical Power Analysis for the Behavioral Sciences*, 2nd ed.; Lawrence Erlbaum Associates: Hillsdale, NJ, USA, 1988.
64. Pearl, R.L.; Puhl, R.M.; Lessard, L.M.; Himmelstein, M.S.; Foster, G.D. Prevalence and correlates of weight bias internalization in weight management: A multinational study. *SSM-Popul. Health* **2021**, *13*, 100755. [CrossRef]
65. Baile, J.I.; González-Calderón, M.J.; Rabito-Alcón, M.F. Obesity bias in the school setting: A brief report. *Children* **2022**, *9*, 1067. [CrossRef]
66. Nutter, S.; Ireland, A.; Alberga, A.S.; Brun, I.; Lefebvre, D.; Hayden, K.A.; Russell-Mayhew, S. Weight bias in educational settings: A systematic review. *Curr. Obes. Rep.* **2019**, *8*, 185–200. [CrossRef]
67. Bacchini, D.; Licenziati, M.R.; Garrasi, A.; Corciulo, N.; Driul, D.; Tanas, R.; Fiumani, P.M.; Di Pietro, E.; Pesce, S.; Crinò, A.; et al. Bullying and victimization in overweight and obese outpatient children and adolescents: An Italian multicentric study. *PLoS ONE* **2015**, *10*, e0142715. [CrossRef]
68. Peguero, A.A.; Shekarkhar, Z. Latino/a student misbehavior and school punishment. *Hisp. J. Behav. Sci.* **2011**, *33*, 54–70. [CrossRef]
69. Attawell, K. *Behind the Numbers: Ending School Violence and Bullying*; United Nations Educational, Scientific and Cultural Organization: Paris, France, 2019.
70. Spear, L.P. The adolescent brain and age-related behavioral manifestations. *Neurosci. Biobehav. Rev.* **2000**, *24*, 417–463. [CrossRef]
71. Steinberg, L. A social neuroscience perspective on adolescent risk-taking. *Dev. Rev.* **2008**, *28*, 78–106. [CrossRef] [PubMed]
72. Clark, O.; Lee, M.M.; Jingree, M.L.; O'Dwyer, E.; Yue, Y.; Marrero, A.; Tamez, M.; Bhupathiraju, S.N.; Mattei, J. Weight stigma and social media: Evidence and public health solutions. *Front. Nutr.* **2021**, *8*, 739056. [CrossRef] [PubMed]
73. Lee, M.; Ata, R.N.; Brannick, M.T. Malleability of weight-biased attitudes and beliefs: A meta-analysis of weight bias reduction interventions. *Body Image* **2014**, *11*, 251–259. [CrossRef] [PubMed]
74. Roberts, C.; Torgerson, D.J. Baseline imbalance in randomised controlled trials. *BMJ* **1999**, *319*, 185. [CrossRef]
75. Sedgwick, P.; Greenwood, N. Understanding the Hawthorne effect. *BMJ* **2015**, *351*, h4672. [CrossRef]
76. Fairburn, C.G.; Cooper, Z. The Eating Disorder Examination. In *Binge Eating: Nature, Assessment, and Treatment*, 12th ed.; Fairburn, C.G., Wilson, G.T., Eds.; Guilford Press: New York, NY, USA, 1993; pp. 317–360.
77. McLean, S.A.; Paxton, S.J. Body image in the context of eating disorders. *Psychiatr. Clin. N. Am.* **2019**, *42*, 145–156. [CrossRef]
78. D'Adamo, L.; Shonrock, A.T.; Monocello, L.; Goldberg, J.; Yaeger, L.H.; Pearl, R.L.; Wilfley, D.E. Psychological interventions for internalized weight stigma: A systematic scoping review of feasibility, acceptability, and preliminary efficacy. *J. Eat. Disord.* **2024**, *12*, 197. [CrossRef]
79. Watts, K.; Cranney, J. The nature and implications of implicit weight bias. *Curr. Psychiatry Rev.* **2009**, *5*, 110–126. [CrossRef]
80. Nutter, S.; Eggerichs, L.A.; Nagpal, T.S.; Salas, X.R.; Chea, C.C.; Saiful, S.; Ralston, J.; Barata-Cavalcanti, O.; Batz, C.; Baur, L.A.; et al. Changing the global obesity narrative to recognize and reduce weight stigma: A position statement from the World Obesity Federation. *Obes. Rev.* **2024**, *25*, e13642. [CrossRef] [PubMed]

children

MDPI

Article

Unlocking Youth Creativity: The Power of Socioemotional Skills

Cátia Branquinho [1,2,3], Catarina Noronha [1], Marina Carvalho [1,4], Nuno Neto Rodrigues [5] and Margarida Gaspar de Matos [1,2,3,6,*]

1 Aventura Social, University of Lisbon, 1400-415 Lisbon, Portugal; cbranquinho@ucp.pt (C.B.); catarinanoronha26@live.com.pt (C.N.); marina.carvalho@ismat.pt (M.C.)
2 Faculdade de Ciências Humanas, University Católica Portuguesa, Palma de Cima, 1649-023 Lisbon, Portugal
3 Institute of Environmental Health (ISAMB), University of Lisbon, 1649-004 Lisbon, Portugal
4 Department of Psychology and Physical Education, Instituto Superior Manuel Teixeira Gomes (ISMAT), 8500-590 Portimão, Portugal
5 Direção Geral de Estatísticas da Educação e Ciência, Avenida 24 de Julho 134, 1200-774 Lisbon, Portugal; nuno.rodrigues@dgeec.medu.pt
6 Applied Psychology Research Centre Capabilities and Inclusion, Instituto Superior de Psicologia Aplicada, 1140-041 Lisbon, Portugal
* Correspondence: mmmatos@ucp.pt

Abstract

Background/Objectives: Creativity has become an essential skill for children and adolescents to cope with the challenges of contemporary society. Beyond academic success, creativity is closely linked to well-being, social adjustment, and personal development. Schools, therefore, play a crucial role in creating conditions that allow students to explore ideas, express themselves, and develop socioemotional resources. This study aimed to examine how self-perceived creativity relates to educational, socioemotional, and well-being factors in Portuguese students, to identify different creativity profiles, and to explore the main variables that predict creativity. Methods: This cross-sectional study was based on secondary analyses of national data from the project Psychological Health and Well-being | School Observatory. The sample included 3011 students aged between 9 and 20 years (M = 13.62; SD = 2.53), from grades 5 to 12. Data were collected using validated instruments: the OECD Socioemotional Skills Survey (SSES), the Positive Youth Development (PYD) scale, and the WHO-5 Well-Being Index. Analyses included group comparisons, cluster analysis to identify self-perceived creativity profiles, correlation analyses, and multiple regression models. Results: Self-perceived creativity did not differ between boys and girls, but it decreased significantly with higher grade levels. Three profiles were identified: low, medium, and high self-perceived creativity. Students with higher self-perceived creativity reported better well-being, more positive relationships with teachers, a stronger sense of belonging at school, and higher parental educational levels. Self-perceived creativity was positively associated with socioemotional skills such as curiosity, sociability, and optimism, as well as with PYD dimensions and well-being. Negative associations were found with age and test anxiety. Socioemotional variables were the strongest predictors of creativity, explaining 39% of its variance. Conclusions: These results show that creativity is closely connected to students' socioemotional development. Investing in emotional skills, supportive relationships, and positive school environments may be a powerful way to foster creativity and promote healthier, more balanced development. This has important implications for educational practice and policy.

Keywords: creativity; socioemotional competences; school-aged children; positive youth development; well-being; age and gender differences

https://doi.org/10.3390/children13020261

1. Introduction

1.1. The Relevance of Creativity in the 21st Century

In the 21st century, there has been widespread recognition of the importance of creativity as a crucial talent for the growth, well-being, and survival of school-aged children. Education plays a vital role in fostering and enhancing creative abilities, especially as the world faces numerous complex challenges that require innovative solutions [1,2].

In educational research, creativity is increasingly understood not only as the production of original and useful outcomes, but also as a developmental and dispositional process grounded in an individual's personally meaningful interpretations of experience and in their self-perceptions of creative potential [3]. In the present study, creativity is conceptualized as self-perceived creativity, assessed through the SSES | Creativity subscale, capturing students' subjective evaluation of their creative tendencies rather than objective creative performance. This approach aligns with a domain-general and socioemotional perspective of creativity, embedded in everyday school functioning rather than domain-specific artistic or academic production.

According to Kupers et al. [2], creativity can be understood as a human characteristic that develops over time through exposure to new ideas and processes, involving collaborative relationships among peers, teachers, and the surrounding environment. The iterative creative process uses feedback from one iteration to inform the next. This perspective supports the view of creativity as a dynamic, socially embedded competence that can be nurtured through educational experiences and socioemotional development.

1.2. Creativity in Educational Contexts

To nurture and support student creativity, a supportive learning environment is essential. In a systematic review, van der Zanden et al. [4] identified four categories of factors related to creativity: individual, parental, educational, and social. Richardson and Mishra [5] assert that learning engagement, the teaching environment, and physical surroundings can all influence students' creative thinking. Every subject has the potential to foster creativity, and as Albar and Southcott [6] demonstrate, the aim of education should be to impart knowledge while also fostering critical thinking skills. In schools, students who receive only direct instruction may become less creative and more passive learners [2]. The pro-creativity pedagogical method, according to Cremin and Chappell [7], comprises seven interconnected features: idea generation and exploration, playfulness, problem-solving, risk-taking, co-construction and collaboration, and teacher creativity. Together, these features emphasize that creativity is not merely an individual attribute, but a pedagogical and relational process shaped by classroom climate, teacher practices, and opportunities for student agency and exploration.

Furthermore, Barbot et al. [8] identify three reasons for differences in originality among young people: task specificity, personal resources, and environmental effects. This framework highlights that creativity emerges from the interaction between individual characteristics (such as cognitive abilities, personality traits, and motivation), the nature of the tasks students engage in, and the quality of the surrounding educational and social context.

According to the literature, schools serve as the main setting for delivering universal early interventions to school-aged children, as nearly all children attend school [9]. Over the past decade, there has been a growing adoption of universal, school-based socioemotional learning programs, aimed at strengthening children's social and emotional skills to support their mental health [10–12]. This reinforces the relevance of studying creativity within a socioemotional framework, as creativity is increasingly viewed as intertwined with emotional regulation, motivation, engagement, and well-being.

https://doi.org/10.3390/children13020261

1.3. Individual and Contextual Factors Associated with Creativity

Regarding age, research identifies developmental "slumps and pumps" contingent upon specific contextual and cultural needs. It has been proposed that as people age and acquire new skills, their creativity may increase [8]. A study by Razumnikova et al. [13] found that young people outperformed children in verbal and figurative creativity as well as fluid intelligence, although age-related variations in creativity depend on the situation. Furthermore, a longitudinal study by Claxton et al. [14] examining the cognitive and affective aspects of creativity in children from the fourth to the ninth grades found that divergent thinking and divergent emotion increase throughout life.

These findings suggest that developmental changes in creativity are not linear and may be reflected both in chronological age and in school grade, which represents a key educational and social context for development. In the present study, grade level is used to capture educational-stage differences, whereas age is considered a continuous developmental indicator. Additionally, studies indicate that individuals with higher creativity levels are better equipped to handle perceived stress in challenging situations. This ability to manage stress effectively is linked to experiencing more positive emotions, leading to the conclusion that creativity is associated with enhanced personal well-being [15].

Overall, socioemotional skills tend to develop exponentially during the transition from childhood to adolescence [16]. These skills will accompany young people throughout their development and are important to the consolidation of their personality and behaviors [16]. The development of socioemotional skills appears in the literature as a promoter of good development and better adaptation of children and young people to the demands inherent to development [17]. This highlights the potential role of socioemotional competencies as central mechanisms through which creativity is expressed and supported.

1.4. Interconnections Between Creativity, Socioemotional Competencies, Well-Being, and Positive Youth Development

The literature suggests that enhancing socioemotional skills can promote creative thinking and self-directed learning, which are essential components of positive youth development [18]. Fostering socioemotional skills—creativity, in particular—contributes to resilience and the ability to cope with challenges, thereby promoting their well-being and supporting positive youth development [19]. From this perspective, creativity can be understood both as an outcome of socioemotional development and as a resource that strengthens adaptive functioning, autonomy, and engagement.

The novelty of the present study lies in the integration of self-perceived creativity with a comprehensive socioemotional framework (SSES), subjective well-being (WHO-5), and Positive Youth Development (PYD), using a large national dataset. In addition, the study adopts a profiling approach that identifies distinct creativity profiles and examines developmental trends across school grades, which remains underexplored in previous research using similar frameworks.

This study aims to examine self-perceived creativity in children and adolescents, investigating its variation by demographic factors such as gender, grade level, and parental education. Additionally, it explores how socioemotional competencies (SSES), well-being (WHO-5) [20], and Positive Youth Development (PYD) resources relate to and predict SSES | Creativity. The analyses follow a sequential approach, starting with descriptive and inferential assessments, then identifying creativity profiles through cluster analysis, and finally conducting regression analyses to determine key predictors of SSES | Creativity.

Based on previous literature, we hypothesize that:

(i) Self-perceived creativity will decline across school grades, reflecting developmental "slumps";

 https://doi.org/10.3390/children13020261

(ii) No substantial gender differences will be observed;
(iii) Socioemotional competencies will emerge as stronger predictors of creativity than demographic variables, well-being, or PYD indicators.

2. Materials and Methods

2.1. Participants

This study included 3011 school-aged children, 1522 girls (50.5%) and 1489 boys (49.5%), aged 9 to 20 years old (M = 13.62; SD = 2.532), from grades 5 to 12 in schools across mainland Portugal. Regarding school level, 25.6% were enrolled in lower secondary education (5th and 6th grades; ages ± 11–12), 40.7% in middle secondary education (7th to 9th grade; ages ± 13–15), and 33.7% in upper secondary education (10th to 12th grade; ages ± 16–18). The vast majority (92.5%) held Portuguese nationality.

With respect to the educational level of the father, 39.1% had completed upper secondary education, 34.2% had completed basic education, 21.6% held a higher education degree, and 5.1% had not completed any level of formal education. Regarding the mother, 38.8% had completed upper secondary education, 34.3% had completed higher education, 23.6% had completed basic education, and 3.4% had not completed any level of formal education. Although schools were selected through a stratified sampling procedure (see Section 2.2), the final sample should be interpreted as not fully representative due to non-proportional stratification, voluntary participation, and missing data. Sample size varied across analyses due to item-level missingness; the effective N is reported for each analysis.

Although the sampling framework was designed to ensure geographical coverage through stratified school selection, the final analytical sample cannot be considered fully representative of the Portuguese student population due to the non-proportional nature of the stratification, voluntary participation at both school and student levels, the absence of sampling weights, and missing data; therefore, representativeness applies to the school selection strategy but not necessarily to the final sample composition. Information regarding the number of schools and classes invited versus those that effectively participated was not systematically recorded, which prevents the calculation of precise response rates. This constitutes an additional limitation of the study.

2.2. Procedure

Under the oversight of the Portuguese Ministry of Education, the research project titled "Psychological Health and Well-being | School Observatory" was created in collaboration with the following organizations: the National Program for the Promotion of School Success, the Portuguese Psychologists Association, the Calouste Gulbenkian Foundation, the Aventura Social Team, and the Directorate-General for Education and Science Statistics.

The sample consisted of a stratified (non-proportional) random selection of public mainland schools, stratified by geographic region/NUTS III (Nomenclature of Territorial Units for Statistics). Data were collected between 23 January and 9 June 2024. Following the stratified random selection of classes by grade level, the teachers and school psychologists responsible for the participating schools administered the data collection instruments in computer labs. Students completed the questionnaires exclusively online, with prior informed consent obtained from their parents or legal guardians. The administration protocol lasted on average between 20 and 30 min. The methods and results are described in detail in the online study report, available at https://info.dgeec.medu.pt/saude-psicologica-e-bem-estar-2024/ (accessed on 10 December 2025) or Matos et al. [21].

Regarding ethics and data protection, data collection was conducted under the authorization and oversight of the Directorate-General for Education and Science Statistics

https://doi.org/10.3390/children13020261

(DGEEC), within the Portuguese Ministry of Education. Participation was voluntary and required informed consent from parents/legal guardians, and student assent when appropriate. Data were collected anonymously online and processed in accordance with the GDPR and the Declaration of Helsinki.

2.3. Measures

The primary goal of the "Psychological Health and Well-being | School Observatory" was to examine the key variables associated with psychological health and well-being in three student age groups: middle school students (self-reports), high school students (self-reports), and preschool and elementary school students (reports by teachers/educators). In addition to student data, self-report information was also collected from teachers, school management staff, educational psychologists, support technicians, operational assistants, and parents or legal guardians, offering a comprehensive perspective on the school ecosystem.

Only data from lower secondary education to upper secondary education were used in the present study. The research protocol included individual variables related to psychological health and well-being, along with sociodemographic questions. Several standardized instruments were employed (Tables 1 and 2).

Table 1. Description of the instruments and variables assessed in the study, including the number of items and response formats.

Variable	Number of Items	Scale
Gender	1	1 = Male and 2 = Female
School grade	1	1 = 5th grade; 2 = 6th grade; 3 = 7th grade; 4 = 8th grade; 5 = 9th grade; 6 = 10th grade; 7 = 11th grade; 8 = 12th grade
Parental educational level	1	1 = no completed school level; 2 = basic education; 3 = upper secondary education; 4 = higher education
SSES | Socioemotional Skills		
Optimism Emotional regulation Resilience Trust Curiosity Sociability Persistence Creativity Energy Cooperation Self-control Sense of belonging to school	8 8 8 8 8 8 8 8 8 8 8 6	0 = Strongly disagree to 4 = Strongly agree
Bullying Relationship with teachers	4 3	0 = Never or almost never to 3 = Once a week or more
Test anxiety	3	0 = Strongly disagree to 4 = Strongly agree
PYD | Positive Youth Development		
Competence Confidence Connection Contribution	6 6 8 5	(used for most items): 0 = Strongly disagree to 4 = Strongly agree (used only for items i and j): 0 = Never true to 4 = Always true 0 = No time to 4 = 6 or more hours
Well-being | WHO-5	5	0 = Never to 5 = All of the time

Note: SSES—Socioemotional Competences Scale; PYD—Positive Youth Development.

https://doi.org/10.3390/children13020261

Table 2. Measurement levels, ranges, descriptive statistics, and reliability of the study variables.

Variable	Level of Measurement	Range	M	SD	Cronbach's Alpha
Gender	Nominal	1–2	1.51	0.5	--
School grade	Ordinal	1–8	4.39	2.23	--
Age	Interval	9–20	13.62	2.53	--
Father's educational level	Ordinal	1–4	2.77	0.84	--
Mother's educational level	Ordinal	1–4	3.04	0.85	--
SSES I Creativity (dependent variable)	Interval	0–4	2.52	0.66	0.83
SSES I Optimism	Interval	0–4	2.71	0.77	0.85
SSES I Emotional regulation	Interval	0–4	2.15	0.73	0.77
SSES I Resilience	Interval	0–4	2.05	0.86	0.88
SSES I Trust	Interval	0–4	2.12	0.73	0.85
SSES I Curiosity	Interval	0–4	2.63	0.69	0.85
SSES I Sociability	Interval	0–4	2.49	0.72	0.82
SSES I Persistence	Interval	0–4	2.59	0.7	0.85
SSES I Energy	Interval	0–4	2.37	0.7	0.81
SSES I Cooperation	Interval	0–4	2.91	0.64	0.88
SSES I Self-control	Interval	0–4	2.5	0.65	0.84
SSES I Sense of belonging to school	Interval	0–4	2.47	0.53	0.76
SSES I Bullying	Interval	0–3	0.44	0.68	0.84
SSES I Relationship with teachers	Interval	0–3	2.32	0.78	0.81
SSES I Test anxiety	Interval	0–4	2.49	1.05	0.85
PYD I Competence	Interval	0–24	14.38	4.7	0.84
PYD I Confidence	Interval	0–24	15.17	5.44	0.9
PYD I Connection	Interval	0–30	18.86	5.7	0.85
PYD I Contribution	Interval	0–20	8.87	4.4	0.72
Well-being I WHO-5	Interval	0–25	15.73	4	0.84

One of the main tools used was the Socioemotional Skills Survey (SSES), developed by the Organisation for Economic Co-operation and Development [22], from which a wide range of variables were drawn. These included the following subscales: SSES I Optimism, Emotional regulation, Resilience, Trust, Confidence, Curiosity, Sociability, Persistence, Creativity, Energy, Cooperation, Self-control, Sense of belonging to school, Bullying, Relationship with teachers, and Test anxiety. For all subscales, higher values indicate higher levels of the respective competence, except for Bullying, where higher scores reflect greater victimization experiences, and Test Anxiety, where higher scores reflect higher levels of anxiety.

The SSES Creativity subscale assesses students' self-perceived tendency to generate new ideas, approach problems in original ways, and engage in imaginative thinking in everyday contexts. All SSES subscales were computed as mean scores across items (range 0–4). Negatively worded items were reverse-coded prior to score computation, following the official OECD scoring guidelines for the SSES instrument.

In addition, the WHO-5 Well-Being Index [20] was applied to assess students' subjective well-being. Finally, the Positive Youth Development (PYD) [23] measure was used, based on the Portuguese-validated version of the instrument [24], ensuring cultural and contextual adequacy. Additionally, the variables of gender, school grade, father's educational level, and mother's educational level were also used.

https://doi.org/10.3390/children13020261

2.4. Statistical Procedures

The data were analyzed using SPSS version 29.0 (SPSS Inc., Chicago, IL, USA). Statistical analysis was conducted in several stages. Initially, the SSES | Creativity subscale was examined descriptively by gender and grade level. Analysis of Variance (ANOVA) revealed significant differences between grade levels, and Tukey's HSD post hoc tests were applied to identify which specific grades differed.

Additionally, we tested the Gender × Grade interaction; as it was not statistically significant, results are presented as main effects.

Next, a non-hierarchical cluster analysis (K-means) was performed using standardized SSES | Creativity scores to group participants into homogeneous profiles. A three-cluster solution was selected to facilitate the interpretation and comparison of low, moderate, and high creativity profiles. Subsequently, clusters were compared on SSES variables, WHO-5, PYD dimensions, and parental education using ANOVAs.

Because clustering was based on a single variable, this procedure is conceptually similar to empirically derived cut-points (e.g., tertiles). This approach was used to support profile comparisons, and its limitations are acknowledged. The clustering approach was preferred over arbitrary cut-off methods (such as tertiles or ±1 SD) because it allows for empirically derived group boundaries based on the actual distribution of creativity scores in the sample.

To characterize and compare the creativity profiles, a one-way ANOVA was applied. Tukey's HSD post hoc tests were used to determine which cluster pairs showed statistically significant differences. When the homogeneity of variances was violated (Levene's test $p < 0.05$), the Games–Howell post hoc test was used because it is robust to unequal variances and unequal group sizes.

Finally, to identify the main predictors of SSES | Creativity, a hierarchical multiple regression was conducted. The model was built in three stages: (i) demographic variables (grade level, gender, and parental education) were introduced; (ii) all SSES were added; (iii) Well-being | WHO-5 and PYD were incorporated.

Spearman's rho correlations were computed because parental education is an ordinal variable, and a preliminary inspection indicated deviations from normality in several distributions. Normality was assessed through the inspection of skewness and kurtosis values and visual examination of histograms.

Missing data were handled using listwise deletion within each analysis; therefore, sample sizes varied across ANOVA, cluster, correlation, and regression models depending on data availability. The effective sample size was $N = 3011$ for descriptive analyses, $N = 2734$ for cluster analyses, and $N = 2517$ for regression models.

All analyses were conducted with a 95% confidence level, ensuring the robustness and reliability of the results.

3. Results

3.1. Gender Differences in SSES | Creativity

ANOVA showed no statistically significant differences between genders on this subscale. Initially, the SSES | Creativity subscale was analyzed according to the participants' gender. The obtained means indicated similar levels of self-perceived creativity between boys ($M = 2.54$; $SD = 0.64$) and girls ($M = 2.51$; $SD = 0.67$).

3.2. SSES | Creativity Across Grade Levels

Subsequently, the variable grade level, ranging from 5th to 12th grades, was examined. Results showed that students in 5th grade had the highest mean score on the SSES | Creativity ($M = 2.76$; $SD = 0.72$), whereas the 9th grade group registered the lowest

 https://doi.org/10.3390/children13020261

mean ($M = 2.4$; $SD = 0.60$). ANOVA revealed statistically significant differences between school grades ($F(7, 2921) = 10.41$, $p < 0.001$, $\eta^2 = 0.024$), suggesting a significant association between school grade and SSES | Creativity. Although statistically significant, this effect size is small, indicating that grade level explains a limited proportion of the variance in self-perceived creativity.

To identify where these differences occurred, Tukey's HSD post hoc tests were conducted. The results showed that students in the 5th grade scored significantly higher than those in the 8th grade (Mdiff = 0.31, $p < 0.05$), 9th grade (Mdiff = 0.36, $p < 0.001$), 10th grade (Mdiff = 0.26, $p < 0.05$), 11th grade (Mdiff = 0.30, $p < 0.05$), and 12th grade (Mdiff = 0.27, $p < 0.05$). No significant differences were found between the 5th and 6th grades.

Similarly, students in the 6th grade showed significantly higher creativity scores than those in the 8th grade (Mdiff = 0.14, $p < 0.05$), 9th grade (Mdiff = 0.19, $p < 0.01$), 11th grade (Mdiff = 0.13, $p < 0.05$), and 12th grade (Mdiff = 0.12, $p < 0.05$).

Taken together, these findings suggest that self-perceived creativity tends to be higher in the early school years, particularly in the 5th and 6th grades, and gradually decreases in later grades, with the lowest values observed in the 9th grade. The remaining grades (7th, 8th, 10th, 11th, and 12th) showed intermediate mean scores, falling between the higher levels of the 5th grade and the lower levels of the 9th grade, which is consistent with the descriptive results presented in Tables 3 and 4.

Table 3. Descriptive statistics for the creativity subscale by grade level.

Grade Level	*N*	*Mean*	*SD*
5th	330	**2.76**	**0.72**
6th	410	2.59	0.70
7th	411	2.55	0.64
8th	374	2.45	0.65
9th	415	2.40	0.60
10th	307	2.50	0.65
11th	371	2.46	0.58
12th	311	2.49	0.64

Note: Bold—highest score.

Table 4. Tukey post hoc test results for the Creativity subscale between grade levels.

Comparison	Mean Difference
5th vs. 8th	0.31 *
5th vs. 9th	0.36 ***
5th vs. 10th	0.26 *
5th vs. 11th	0.3 *
5th vs. 12th	0.27 *
6th vs. 8th	0.14 *
6th vs. 9th	0.19 **
6th vs. 11th	0.13 *
6th vs. 12th	0.12 *

Note: * $p < 0.05$, ** $p < 0.01$, *** $p < 0.001$.

3.3. Identification of SSES | Creativity Profiles: Cluster Analysis

To identify homogeneous groups of participants, a non-hierarchical cluster analysis (K-means) was conducted based on the SSES | Creativity subscale. The final clusters, using standardized scores (Z-scores) from the SSES | Creativity subscale, were as follows:

- Cluster 1 (Low Creativity): $Z = -1.456$
- Cluster 2 (High Creativity): $Z = 1.168$

https://doi.org/10.3390/children13020261

- Cluster 3 (Moderate Creativity): $Z = -0.290$

These clusters represent distinct creativity profiles: Cluster 1 includes participants with scores significantly below the sample mean ($n = 1064$), Cluster 2 includes those with scores significantly above the mean ($n = 461$), and Cluster 3 comprises individuals whose scores are close to the overall sample average ($n = 1209$).

Before examining differences between creativity profiles across socioemotional, well-being, and developmental variables, a one-way ANOVA was conducted to confirm that the identified clusters differed significantly in creativity. Results revealed a significant effect of cluster membership on creativity, $F(2, 2731) = 5706.73$, $p < 0.001$, with a very large effect size ($\eta^2 = 0.81$). This represents an extremely large effect size, confirming the strong differentiation between the three creativity profiles. Post hoc comparisons using the Games–Howell test indicated that all three clusters differed significantly from each other ($p < 0.001$), supporting the distinctiveness of the low, moderate, and high creativity profiles.

The next step involved conducting Analyses of Variance (ANOVAs) to compare the mean scores of the clusters across various SSES. The ANOVA results revealed significant differences between the clusters for all variables assessed ($p < 0.05$). Tukey HSD post hoc tests showed the following patterns:

- Cluster 2 (High Creativity) consistently showed the highest mean scores across almost all SSES indicators, Well-being | WHO-5, and PYD variables, except for SSES | Bullying and SSES | Test anxiety, with significant differences from the other clusters ($p < 0.001$).
- Cluster 1 (Low Creativity) had the lowest mean scores on most SSES indicators, Well-being | WHO-5, and PYD, with significant differences from Clusters 2 and 3 ($p < 0.05$), except for a few variables.
- Cluster 3 (Moderate Creativity) fell between the other clusters, with mean scores significantly higher than Cluster 1 and lower than Cluster 2 for most variables ($p < 0.05$).

Notable exceptions were observed in a few specific variables. In terms of SSES | Resilience/stress resistance, no significant difference was found between Clusters 1 and 3; however, both differed significantly from Cluster 2, which exhibited higher resilience levels. Regarding SSES | Bullying, Cluster 2 demonstrated significantly lower levels of victimization compared to Clusters 1 and 3, between which no significant differences emerged. For SSES | Test anxiety, participants in Cluster 1 reported significantly higher anxiety than those in Cluster 2, while no significant differences were found between Clusters 1 and 3. Finally, in relation to Parents' educational level, only Cluster 2 stood out, presenting significantly higher values than Clusters 1 and 3, which did not differ from one another (Table 5).

Table 5. Descriptive statistics and ANOVA results by cluster.

Variable	Cluster 1 (*M*, *SD*)	Cluster 2 (*M*, *SD*)	Cluster 3 (*M*, *SD*)	*F*	η^2	Post Hoc (Tukey HSD)
SSES | Optimism	2.25 (0.83)	**3.03 (0.71)**	2.66 (0.7)	183.27 ***	0.112	C2 > C3 > C1
SSES | Emotional regulation	1.88 (0.74)	**2.36 (0.81)**	2.11 (0.64)	70.2 ***	0.046	C2 > C3 > C1
SSES | Resilience	1.89 (0.92)	**2.23 (0.91)**	1.99 (0.78)	32.71 ***	0.022	C2 > (C1 = C3)
SSES | Trust	1.83 (0.72)	**2.29 (0.8)**	2.11 (0.65)	62.31 ***	0.041	C2 > C3 > C1
SSES | Curiosity	2.13 (0.77)	**3.03 (0.62)**	2.54 (0.57)	353.07 ***	0.195	C2 > C3 > C1
SSES | Sociability	2.05 (0.75)	**2.79 (0.74)**	2.43 (0.61)	185.6 ***	0.113	C2 > C3 > C1
SSES | Persistence	2.18 (0.72)	**2.95 (0.69)**	2.50 (0.59)	245.96 ***	0.144	C2 > C3 > C1
SSES | Creativity	1.91 (0.65)	**2.76 (0.74)**	2.27 (0.56)	294.97 ***	0.169	C2 > C3 > C1
SSES | Energy	2.57 (0.78)	**3.25 (0.55)**	2.81 (0.56)	242.55 ***	0.143	C2 > C3 > C1
SSES | Cooperation	2.18 (0.70)	**2.78 (0.67)**	2.43 (0.55)	166.3 ***	0.103	C2 > C3 > C1
SSES | Self-control	2.26 (0.52)	**2.66 (0.57)**	2.42 (0.47)	105.64 ***	0.068	C2 > C3 > C1
SSES | Bullying	0.48 (0.65)	0.37 (0.59)	0.48 (0.71)	8.65 ***	0.006	C2 < (C1 = C3)

https://doi.org/10.3390/children13020261

Table 5. *Cont.*

Variable	Cluster 1 (*M, SD*)	Cluster 2 (*M, SD*)	Cluster 3 (*M, SD*)	*F*	η^2	Post Hoc (Tukey HSD)
SSES I Relationship with teachers	2.14 (0.88)	**2.49 (0.7)**	2.28 (0.77)	36.63 ***	0.025	C2 > C3 > C1
SSES I Test anxiety	**2.61 (1.13)**	2.41 (1.14)	2.50 (0.97)	5.36 **	0.004	C1 > C2, (C1 = C3), (C2 = C3)
PYD I Competence	11.61 (5.09)	**16.26 (4.5)**	14.05 (4.22)	168.74 ***	0.104	C2 > C3 > C1
PYD I Confidence	11.87 (5.72)	**17.46 (5.29)**	14.74 (4.84)	185.59 ***	0.113	C2 > C3 > C1
PYD I Connection	15.63 (6.16)	**21.07 (5.66)**	18.51 (4.96)	157.95 ***	0.098	C2 > C3 > C1
PYD I Contribution	7.64 (4.46)	**9.67 (4.22)**	8.75 (4.39)	32.9 ***	0.022	C2 > C3 > C1
Well-being I WHO-5	13.83 (5.25)	**17.06 (4.75)**	15.45 (4.86)	66.9 ***	0.045	C2 > C3 > C1
Father's educational level	2.67 (0.81)	**2.94 (0.84)**	2.7 (0.83)	26.64 ***	0.019	C2 > (C1 = C3)
Mother's educational level	2.97 (0.86)	**3.22 (0.81)**	2.95 (0.85)	30.67 ***	0.021	C2 > (C1 = C3)

Note: Bold—higher mean; ** $p < 0.01$, *** $p < 0.001$.

Based on these results, the competencies associated with each creativity profile can be summarized as follows (Figure 1):

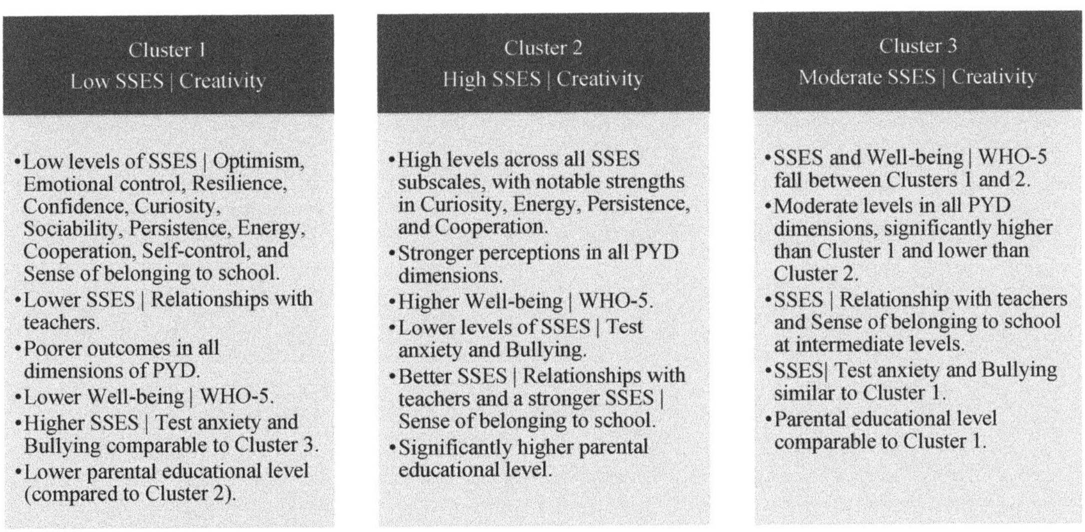

Figure 1. Conceptual summary of the low, moderate, and high creativity profiles.

3.4. Correlations with SSES I Creativity

School grade was used as the primary developmental indicator in group comparisons because it reflects students' educational context, whereas age was included in correlational analyses as a continuous developmental variable.

Correlational analysis was subsequently conducted to examine the direction and magnitude of the associations between SSES I Creativity and the other study variables, serving as the statistical basis for the hierarchical regression that followed.

This analysis revealed a significant negative correlation between SSES I Creativity and age ($\rho = -0.107$, $p < 0.001$), indicating that younger students tend to report higher levels of perceived creativity.

Regarding parental education level, significant positive correlations were observed with SSES I Creativity for both the father's level of education ($\rho = 0.134$, $p < 0.001$) and that of the mother ($\rho = 0.129$, $p < 0.001$). This suggests that a higher level of parental education is associated with higher levels of creativity in students.

 https://doi.org/10.3390/children13020261

The SSES subscales showed strong relationships with SSES | Creativity, with the most notable being SSES | Curiosity ($\rho = 0.507$, $p < 0.001$), SSES | Energy ($\rho = 0.458$, $p < 0.001$), SSES | Cooperation ($\rho = 0.459$, $p < 0.001$), SSES | Persistence ($\rho = 0.447$, $p < 0.001$), SSES | Optimism ($\rho = 0.368$, $p < 0.001$), SSES | Sociability ($\rho = 0.394$, $p < 0.001$), SSES | Emotional regulation ($\rho = 0.251$, $p < 0.001$), and SSES | Resilience ($\rho = 0.153$, $p < 0.001$).

Additionally, SSES | Creativity was positively correlated with SSES | Sense of belonging to school ($\rho = 0.319$, $p < 0.001$), SSES | Relationship with teachers ($\rho = 0.192$, $p < 0.001$), SSES | Self-control ($\rho = 0.384$, $p < 0.001$), and well-being measured by the Well-being | WHO-5 ($\rho = 0.234$, $p < 0.001$). Conversely, a negative correlation was found between SSES | Creativity and SSES | Bullying ($\rho = -0.091$, $p < 0.001$), while a non-significant negative correlation was observed with SSES | Test anxiety.

Within the PYD framework, SSES | Creativity showed significant positive associations with the dimensions of PYD | Competence ($\rho = 0.373$, $p < 0.001$), PYD | Confidence ($\rho = 0.387$, $p < 0.001$), PYD | Connection ($\rho = 0.366$, $p < 0.001$), and PYD | Contribution ($\rho = 0.177$, $p < 0.001$).

3.5. Predictors of SSES | Creativity: Hierarchical Multiple Regression

A hierarchical multiple regression was conducted to assess the predictive capacity of demographic variables, SSES, PYD, and Well-being | WHO-5 on the dependent variable SSES | Creativity.

In Model 1, demographic variables (grade level, gender, and parental education) were entered as control variables. This model was statistically significant ($F(4, 2512) = 20.897$, $p < 0.001$), explained only 3.2% of the total variance in SSES | Creativity ($R^2 < 0.05$). Within this block, both grade level and parental education significantly predicted SSES | Creativity. Grade level showed a negative association with SSES | Creativity ($\beta = -0.092$, $p < 0.001$), whereas parental education was positively associated (father's education: $\beta = 0.07$, $p < 0.01$; mother's education: $\beta = 0.091$, $p < 0.001$). Gender did not emerge as a significant predictor of SSES | Creativity.

In Model 2, the SSES subscales were added, resulting in a substantial and statistically significant increase in explained variance ($\Delta R^2 = 0.359$, $p < 0.001$). The full model accounted for 39.1% of the variance in SSES | Creativity ($F(18, 2498) = 89.168$, $p < 0.001$). Examination of standardized coefficients revealed several SSES as robust predictors of SSES | Creativity, even after controlling for demographic variables. The strongest positive predictors were SSES | Curiosity ($\beta = 0.26$, $p < 0.001$), SSES | Energy ($\beta = 0.193$, $p < 0.001$), SSES | Cooperation ($\beta = 0.178$, $p < 0.001$), and SSES | Sociability ($\beta = 0.11$, $p < 0.001$). Conversely, SSES | Trust ($\beta = -0.100$, $p < 0.001$) and SSES | Test anxiety ($\beta = -0.052$, $p < 0.01$) were significant negative predictors. The inclusion of these socioemotional variables diminished the predictive power of most demographic variables, with grade level and father's education level ceasing to be significant predictors.

Finally, Model 3 incorporated Well-being | WHO-5 and PYD variables. However, this addition did not lead to a significant increase in explained variance, indicating that Well-being | WHO-5 and PYD factors did not uniquely contribute to predicting SSES | Creativity beyond the socioemotional competencies already accounted for. Only mother's education level remained positively and significantly associated with SSES | Creativity ($\beta = 0.04$, $p < 0.05$) (Tables 6 and 7).

 https://doi.org/10.3390/children13020261

Table 6. Summary of hierarchical regression models predicting Creativity.

Model	R	R^2	Adjusted R^2	ΔR^2	F Change	df1	df2
1	0.179	0.032	0.031 ***	0.032	20.897	4	2512
2	0.625	0.391	0.387 ***	0.359	105.207	14	2498
3	0.627	0.393	0.387	0.002	1.487	5	2493

Note: $N = 2517$. Model 1 includes grade level, gender, and parental education level. Model 2 adds SSES subscales. Model 3 adds Well-being | WHO-5 and all PYD dimensions. *** $p < 0.001$.

Table 7. Standardized coefficients of hierarchical regression predicting Creativity.

Variable	*B*	*SE*	*β*	*t*
Model 1				
(Constant)	2.315 ***	0.072		31.93
Grade level	−0.027 ***	0.006	−0.092	−4.643
Gender	−0.018	0.026	−0.014	−0.711
Father's educational level	0.054 **	0.019	0.07	2.868
Mother's education level	0.07 ***	0.019	0.091	3.736
Model 2				
(Constant)	0.557 ***	0.099		5.624
Grade level	−0.005	0.005	−0.019	−1.076
Gender	−0.01	0.024	−0.007	−0.409
Father's educational level	0.02	0.015	0.025	1.294
Mother's educational level	0.031 *	0.015	0.041	2.091
SSES | Optimism	0.019	0.02	0.022	0.942
SSES | Emotional regulation	−0.019	0.021	−0.022	−0.94
SSES | Resilience	0.015	0.018	0.02	0.844
SSES | Trust	−0.091 ***	0.018	−0.1	−5.041
SSES | Curiosity	0.249 ***	0.02	0.26	12.28
SSES | Sociability	0.1 ***	0.022	0.11	4.64
SSES | Persistence	0.084 ***	0.02	0.089	4.162
SSES | Energy	0.18 ***	0.021	0.193	8.442
SSES | Cooperation	0.183 ***	0.025	0.178	7.447
SSES | Self-control	0.056 *	0.023	0.055	2.454
SSES | Sense of belonging to school	−0.039	0.028	−0.031	−1.381
SSES | Bullying	−0.006	0.018	−0.005	−0.314
SSES | Relationship with teachers	−0.013	0.015	−0.016	−0.909
SSES | Test anxiety	−0.032 **	0.012	−0.052	−2.735
Model 3				
(Constant)	0.609 ***	0.102		5.995
Mother's education level	0.031 *	0.015	0.04	2.069
SSES | Trust	−0.092 ***	0.019	−0.101	−4.888
SSES | Curiosity	0.249 ***	0.02	0.260	12.242
SSES | Sociability	0.093 ***	0.022	0.102	4.133
SSES | Persistence	0.08 ***	0.020	0.086	3.987
SSES | Energy	0.173 ***	0.023	0.186	7.633
SSES | Cooperation	0.174 ***	0.025	0.169	6.977
SSES | Self-control	0.053 *	0.023	0.052	2.304
SSES | Test anxiety	−0.032 **	0.012	−0.052	−2.704
Well-being | WHO-5	−0.004	0.003	−0.033	−1.566
PYD | Competence	0.004	0.004	0.027	1.004
PYD | Confidence	0.004	0.003	0.033	1.216
PYD | Connection	0.001	0.003	0.005	0.195
PYD | Contribution	0.002	0.003	0.012	0.691

Note: Significant variables with $p < 0.05$ are presented in Models 2 and 3. The non-significant variables from Model 3 (except for those added in this step) were omitted for clarity. * $p < 0.05$; ** $p < 0.01$; *** $p < 0.001$.

https://doi.org/10.3390/children13020261

4. Discussion

The results of this study reveal that SSES | Creativity does not show significant variations between genders, in line with the results of Ogurlu et al. [25] and Nakano et al. [26]. The latter concluded that there is no consensus on gender differences, suggesting that aspects such as opportunities, motivation, and attitudes should be considered a priority. This finding reinforces the idea that creativity-related self-perceptions are not inherently gendered, but rather shaped by contextual and educational opportunities.

With regard to grade level, there is a statistically significant but small decline in SSES | Creativity as students progress through their academic career, especially between 5th and 9th grades. Although statistically significant, the magnitude of this effect is modest ($\eta^2 = 0.024$), suggesting that the observed decline reflects subtle developmental changes rather than a marked loss of creative capacity. This pattern may reflect a global trend of declining skills such as curiosity, creativity, and openness to experience in early adolescence [27,28]. This result is consistent with the literature that identifies creative "slumps" at this stage [7], attributed to cognitive, emotional, and social changes typical of adolescence, although some studies point to creative "pumps" associated with the acquisition of new skills and life experiences. In the present study, this potential increase may be related to the age of entry into secondary school.

This decline highlights a critical window for intervention, underscoring the importance of pedagogical strategies that preserve and stimulate creativity from the age of 11–12. Rather than indicating a disappearance of creativity, this pattern may reflect changes in school demands, evaluation practices, and motivational climate that progressively constrain students' opportunities for creative expression. Although the literature points to the positive effects of implementing creativity promotion programs in schools [29,30], their effective integration continues to face significant barriers, such as curriculum overload, the absence of creativity assessment indicators, and insufficient teacher training and capacity building [31]. Models such as the "7Cs of Creativity" [32] and the "4C Model" [3] offer solid theoretical frameworks to guide these practices, emphasizing the multidimensional and developmental nature of creativity and the need for intentionally structured educational contexts to enhance it. In this sense, the study by Li et al. [33] highlights the emerging role of digital technologies as promising resources to support creative development, especially in interactive learning environments, paving the way for educational practices more aligned with the challenges of the 21st century.

The regression analysis also revealed that dimensions such as SSES | Curiosity, Energy, and Cooperation are the most robust predictors of SSES | Creativity, confirming the relational and developmental nature of this competency. SSES | Creativity should not be seen as a fixed trait, but rather as a dynamic capacity, influenced by other SSES subscales and susceptible to being enhanced through deliberate pedagogical practices [34]. From a developmental and social-cognitive perspective, curiosity reflects intrinsic motivation and openness to experience, which are central mechanisms for creative exploration and idea generation. Energy may be interpreted as an indicator of engagement and vitality that sustains creative effort, whereas cooperation highlights the fundamentally social nature of creativity in school contexts, where ideas are often co-constructed through interaction. These findings are consistent with cognitive approaches to creativity that highlight the contribution of higher-order cognitive processes, including mental imagery and the manipulation of mental representations, to creative expression. Empirical studies have shown that such cognitive resources are associated with creative performance, suggesting that socioemotional competencies such as curiosity and engagement may support the cognitive mechanisms involved in everyday creativity [35–37]. In fact, studies show a significant association between curiosity and creativity [38], reinforcing the relevance of educational

 https://doi.org/10.3390/children13020261

contexts that value exploration, questioning, and reflection. The role of energy/vitality as a creative facilitator [39,40] and cooperation as a catalyst for collective creativity [32,41] reinforces the need for learning strategies that integrate intrinsic motivation, collaborative work, and real-world challenges.

In addition, the data show that SSES | Creativity is positively associated with Well-being | WHO-5 and other socioemotional skills, such as SSES | Emotional regulation, Resilience, and Sense of belonging to school. Students with higher levels of SSES | Creativity perceive themselves as more resilient, emotionally balanced, and integrated into the school context. This result reinforces the literature that associates creativity with positive emotions, reduced perceived stress [15], and greater adaptability [42,43]. The link found between SSES | Creativity and PYD indicators supports the idea that promoting creative skills contributes not only to academic performance but also to the construction of positive identities, greater autonomy, and better social adaptation [19]. Carvalho et al. [44] also showed that adolescents with higher levels of PYD | Competence, Confidence, and Connection reported fewer psychological symptoms and greater life satisfaction, highlighting how PYD dimensions are closely related to well-being outcomes. The literature also shows that SSES | Creativity and Resilience share traits such as flexibility, problem-solving skills, and adaptability [45,46]. However, the absence of a unique contribution of WHO-5 and PYD in the final regression model suggests that their relationship with creativity is largely explained by shared variance with socioemotional skills, indicating that SSES subscales capture core psychological processes linking creativity, well-being, and positive development. The role of the family also emerges as a determining factor in students' creative profile, with parental involvement being fundamental in stimulating curiosity, autonomy, and emotional expression.

In short, the data from this study point to the need for a systemic and integrated approach to promoting creativity, linking schools, families, and communities. Promoting SSES | Creativity involves investing in the cross-curricular development of socioemotional skills and student well-being, creating consistent and equitable learning opportunities throughout the school career. Integrating creativity into curricula, in line with the "4Cs" of 21st-century learning (Critical Thinking, Communication, Collaboration, and Creativity), can not only enhance academic performance [47] but also prepare young people to face the complex challenges of the present and the future in an innovative, ethical, and resilient way [48].

4.1. Strengths and Limitations

The study benefits from a large sample size, which enhances the reliability and generalizability of the results, although the sample is not representative of the population. Its methodological approach is comprehensive and robust, combining cluster analyses and hierarchical multiple regression. This combination allowed for not only the identification of SSES | Creativity profiles but also their validation and the effective determination of their predictors. Furthermore, the variety of variables examined—SSES, Well-being (WHO-5), and PYD—provides a rich and multifaceted context for interpreting the results.

The main limitation lies in the cross-sectional design, which prevents causal inference. The observed decline in SSES | Creativity across advancing school years, for example, is a comparative observation rather than a true longitudinal development. Additionally, SSES | Creativity was assessed through self-report, which may not correspond to objective creativity. Finally, it is important to note that although some associations (such as with education level) were statistically significant, their effect sizes were modest, indicating a limited influence on the variance in creativity. Moreover, students were nested within classes and schools, which may violate the assumption of independence of observations.

https://doi.org/10.3390/children13020261

As multilevel modeling was not applied, potential clustering effects could not be statistically controlled. The use of a one-variable cluster solution also limits the conceptual richness of the profiles, which should be interpreted as empirical groupings rather than multidimensional typologies. Moreover, the large number of statistical tests increases the risk of Type I error, and the substantial reduction in sample size across analyses due to missing data may have affected statistical power and generalizability.

4.2. Implications for the Future and Public Policy

- Based on the present findings, school-based programs targeting socioemotional skills appear to be promising avenues for supporting the development of SSES | Creativity from the early years of elementary education, particularly before the 7th grade. These programs may benefit from curricular approaches that prioritize methodologies fostering curiosity, cooperation, and active engagement.
- Educational policies should support teacher training initiatives that strengthen key socioemotional skills related to creativity, such as curiosity, energy, cooperation, emotional regulation, resilience, persistence, and sense of belonging to school, while fostering safe school environments that value emotional well-being, psychological safety, and positive relationships.
- Family involvement should be encouraged, particularly in contexts of lower parental educational levels, to strengthen children's cultural and social capital. In addition, monitoring systems for SSES | Creativity and related competencies may help tailor interventions and support students' holistic development.

5. Conclusions

This study highlights creativity as a socioemotional and developmental competence that is strongly embedded in students' everyday school experiences. Rather than being primarily shaped by demographic characteristics, self-perceived creativity appears to be closely linked to socioemotional skills such as curiosity, energy, and cooperation, reinforcing the idea that creativity emerges in relational and motivational contexts.

Although a small decline in creativity was observed across grade levels, particularly between the 5th and 9th grades, this finding should be interpreted as a subtle developmental shift rather than a loss of creative potential. It points to early adolescence as a sensitive period in which educational environments may play a crucial role in sustaining students' creative engagement. Overall, the results support an understanding of creativity as a dynamic capacity that can be nurtured through educational practices that promote socioemotional development, well-being, and positive relationships within the school community. Future research using longitudinal and multilevel designs will be essential to deepen the understanding of how creativity evolves over time and across educational contexts.

Key Messages

(1) Self-perceived creativity shows a small but significant decline between 5th and 9th grades, identifying early adolescence as a sensitive period for educational support.
(2) Socioemotional skills, particularly curiosity, energy, and cooperation, are the strongest predictors of creativity.
(3) Creativity is closely linked to students' well-being and positive development, reinforcing its relevance for holistic education.

https://doi.org/10.3390/children13020261

Author Contributions: Conceptualization, C.B. and M.G.d.M.; methodology, C.B. and M.G.d.M.; software, C.B.; validation, M.C., N.N.R. and M.G.d.M.; formal analysis, C.B. and M.G.d.M.; investigation, C.B., C.N. and M.G.d.M.; resources, N.N.R. and M.G.d.M.; data curation, N.N.R. and M.G.d.M.; writing—original draft preparation, C.B. and C.N.; writing—review and editing, C.B.; visualization, C.B. and M.G.d.M.; supervision, M.G.d.M.; project administration, N.N.R. and M.G.d.M.; funding acquisition, N.N.R. and M.G.d.M. All authors have read and agreed to the published version of the manuscript.

Funding: This research was funded by the Direção Geral de Estatísticas da Educação e Ciência (2023/001A/DGEEC/CPrev).

Institutional Review Board Statement: Data collection was conducted under the authorization, oversight, and formal approval of the Consultative Council of the Direção-Geral de Estatísticas da Educação e Ciência (DGEEC), within the framework of the Ministério da Educação, the competent national authority for educational research involving minors. The study was developed in the context of the Observatório da Saúde Psicológica e do Bem-Estar (https://www.dgeec.medu.pt/p/educacao-pre-escolar-basico-e-secundario/observatorios/observatorio-da-saude-psicologica-e-do-bem-estar, accessed on 10 December 2025). Within this authorization process, no formal ethical approval code or protocol number is issued, as the evaluation and approval are conducted through administrative and institutional procedures, including assessment by the Consultative Council, rather than through a numbered ethics committee protocol. The study involved non-invasive and observational data collection using anonymous questionnaires, posed minimal risk to participants, and did not include any clinical, experimental, or therapeutic intervention. The participation by children occurred only after informed consent was obtained from parents or legal guardians, and assent was obtained from the children when appropriate. No identifiable personal data were collected. All procedures complied with national regulations governing educational research and adhered to the ethical principles of the Declaration of Helsinki.

Informed Consent Statement: Informed consent was obtained from all subjects involved in the study.

Data Availability Statement: The datasets generated and/or analyzed during the current study are not publicly available but are available from the corresponding author upon reasonable request.

Acknowledgments: The authors thank the participating schools for their collaboration and institutional support, as well as all children, parents, and other respondents who generously contributed their time and participation to this study.

Conflicts of Interest: The authors declare no conflicts of interest.

References

1. Alves-Oliveira, P.; Arriaga, P.; Xavier, C.; Hoffman, G.; Paiva, A. Creativity Landscapes: Systematic Review Spanning 70 Years of Creativity Interventions for Children. *J. Creat. Behav.* **2022**, *56*, 16–40. [CrossRef]
2. Kupers, E.; Lehmann-Wermser, A.; McPherson, G.; van Geert, P. Children's Creativity: A Theoretical Framework and Systematic Review. *Rev. Educ. Res.* **2019**, *89*, 93–124. [CrossRef]
3. Kaufman, J.C.; Beghetto, R.A. Beyond big and little: The four C model of creativity. *Rev. Gen. Psychol.* **2009**, *13*, 1–12. [CrossRef]
4. van der Zanden, P.; Meijer, P.C.; Beghetto, R.A. A review study about creativity in adolescence: Where is the social context? *Think. Ski. Creat.* **2020**, *38*, 100702. [CrossRef]
5. Richardson, C.; Mishra, P. Learning environments that support student creativity: Developing the SCALE. *Think. Ski. Creat.* **2018**, *27*, 45–54. [CrossRef]
6. Albar, S.B.; Southcott, J.E. Problem and project-based learning through an investigation lesson: Significant gains in creative thinking behaviour within the Australian foundation (preparatory) classroom. *Think. Ski. Creat.* **2021**, *41*, 100853. [CrossRef]
7. Cremin, T.; Chappell, K. Creative pedagogies: A systematic review. *Res. Pap. Educ.* **2021**, *36*, 299–331. [CrossRef]
8. Barbot, B.; Lubart, T.I.; Besançon, M. "Peaks, Slumps, and Bumps": Individual Differences in the Development of Creativity in Children and Adolescents. *New Dir. Child Adolesc. Dev.* **2016**, *2016*, 33–45. [CrossRef] [PubMed]
9. Graham, A. *Early Intervention: The Next Steps: An Independent Report to Her Majesty's Government*; Department for Education: London, UK, 2011. Available online: https://assets.publishing.service.gov.uk/media/5a7b837d40f0b645ba3c4d09/early-intervention-next-steps2.pdf (accessed on 8 December 2025).

 https://doi.org/10.3390/children13020261

10. Cefai, C.; Camilleri, L.; Bartolo, P.; Grazzani, I.; Cavioni, V.; Conte, E.; Ornaghi, V.; Agliati, A.; Gandellini, S.; Tatalovic Vorkapic, S.; et al. The effectiveness of a school-based, universal mental health programme in six European countries. *Front. Psychol.* **2022**, *13*, 925614. [CrossRef] [PubMed]
11. Cipriano, C.; Strambler, M.J.; Naples, L.H.; Ha, C.; Kirk, M.; Wood, M.; Sehgal, K.; Zieher, A.K.; Eveleigh, A.; McCarthy, M.; et al. The state of evidence for social and emotional learning: A contemporary meta-analysis of universal school-based SEL interventions. *Child Dev.* **2023**, *94*, 1181–1204. [CrossRef]
12. Connolly, P.; Miller, S.; Hanratty, J.; Roberts, J.; Sloan, S. Protocol for a systematic review: Universal school-based programmes for improving social and emotional outcomes in children aged 3–11 years: A systematic review and meta-analysis. *Campbell Syst. Rev.* **2018**, *14*, 1–39. [CrossRef] [PubMed]
13. Razumnikova, O.; Bakaev, M. Age-Related Changes and Reorganization of Creativity and Intelligence Indices in Schoolchildren and University Students. *J. Intell.* **2022**, *10*, 52. [CrossRef] [PubMed]
14. Claxton, A.; Pannells, T.; Rhodes, P. Developmental Trends in the Creativity of School-Age Children. *Creat. Res. J.* **2005**, *17*, 327–335. [CrossRef]
15. Fiori, M.; Fischer, S.; Barabasch, A. Creativity is associated with higher well-being and more positive COVID-19 experience. *Pers. Individ. Differ.* **2022**, *194*, 111646. [CrossRef] [PubMed]
16. Schonert-Reichl, K.A.; Lawlor, M.S. The Effects of a Mindfulness-Based Education Program on Pre- and Early Adolescents' Well-being and Social and Emotional Competence. *Mindfulness* **2010**, *1*, 137–151. [CrossRef]
17. Greenberg, M.T.; Weissberg, R.P.; O'Brien, M.U.; Zins, J.E.; Fredericks, L.; Resnik, H.; Elias, M.J. Enhancing school-based prevention and youth development through coordinated social, emotional, and academic learning. *Am. Psychol.* **2003**, *58*, 466–474. [CrossRef]
18. Saleem, S.; Burns, S.; Perlman, M. Cultivating young minds: Exploring the relationship between child socio-emotional competence, early childhood education and care quality, creativity and self-directed learning. *Learn. Individ. Differ.* **2024**, *111*, 102440. [CrossRef]
19. Moreno, A.; Jurado, M.D.M. Social skills and creativity as elements that enhance resilience in adolescence. *Behav. Sci.* **2024**, *14*, 1158. [CrossRef]
20. World Health Organization. *The World Health Organization-Five Well-Being Index (WHO-5)*; World Health Organization: Geneva, Switzerland, 2024. Available online: https://creativecommons.org/licenses/by-nc-sa/3.0/igo/ (accessed on 8 December 2025).
21. Matos, M.G.; Branquinho, C.; Tomé, G.; Santos, O.; Noronha, C.; Moraes, B.; Domingos, L.; Raimundo, M.; Cerqueira, A.; Gaspar, T.; et al. *Relatório Saúde Psicológicar [Psychological Health and Well-Being Report]*; DGEEC: Lisbon, Portugal; Aventura Social Associação: Lisbon, Portugal, 2024.
22. OECD. *Beyond Academic Learning: First Results from the Survey of Social and Emotional Skills*; OECD Publishing: Paris, France, 2021. [CrossRef]
23. Geldhof, G.J.; Bowers, E.P.; Boyd, M.J.; Mueller, M.K.; Napolitano, C.M.; Schmid, K.L.; Lerner, J.V.; Lerner, R.M. Creation of short and very short measures of the five Cs of positive youth development. *J. Res. Adolesc.* **2014**, *24*, 163–176. [CrossRef]
24. Tomé, G.; Matos, M.G.; Camacho, I.; Gomes, P.; Reis, M.; Branquinho, C.; Gomez-Baya, D.; Wiium, N. Positive youth development (pyd-sf): Validação para os adolescentes portugueses. *Psicol. Saúde Doenças* **2019**, *20*, 556–568. [CrossRef]
25. Ogurlu, U.; Yalin, H.S.; Yavuz Birben, F. The Relationship Between Psychological Symptoms, Creativity, and Loneliness in Gifted Children. *J. Educ. Gift.* **2018**, *41*, 193–210. [CrossRef]
26. de Cássia Nakano, T.; da Silva Oliveira, K.; Zaia, P. Gender Differences in Creativity: A Systematic Literature Review. *Psicol. Teor. Pesqui.* **2021**, *37*, e372116. [CrossRef]
27. Engel, S. *The Hungry Mind: The Origins of Curiosity in Childhood*; Harvard University Press: Cambridge, MA, USA, 2015.
28. Heintz, S.; Ruch, W. Cross-sectional age differences in 24 character strengths: Five meta-analyses from early adolescence to late adulthood. *J. Posit. Psychol.* **2022**, *17*, 356–374. [CrossRef]
29. Muñoz Silva, F.D.; Luna Guevara, J.R.; López Regalado, O. Creative thinking in the educational context. *Rev. Cient. UCSA* **2021**, *8*, 39–50. [CrossRef]
30. Rodrigues, M.S.B.; Chagas-Ferreira, J.F. Programs to stimulate creativity in schools: A systematic review. *Linhas Crit.* **2023**, *29*, e47206. [CrossRef]
31. OECD. *Curriculum Overload: A Way Forward*; OECD: Paris, France, 2020.
32. Cropley, D.H.; Kaufman, J.C.; Cropley, A.J. Measuring Creativity for Innovation Management. *J. Technol. Manag. Innov.* **2011**, *6*, 13–30. [CrossRef]
33. Li, Y.; Kim, M.; Palkar, J. Using emerging technologies to promote creativity in education: A systematic review. *Int. J. Educ. Res. Open* **2022**, *3*, 100177. [CrossRef]
34. Vuichard, A.; Capron Puozzo, I. When creativity wakes up emotions: An illustration of a creative course for pre-service teacher training. *Discov. Educ.* **2024**, *3*, 214. [CrossRef]
35. Lu, R.; Zhang, Y.; Bao, N.; Su, M.; Zhang, X.; Shi, J. Visuospatial, rather than verbal working memory capacity plays a key role in verbal and figural creativity. *Think. Reason.* **2022**, *28*, 29–60. [CrossRef]

https://doi.org/10.3390/children13020261

36. Wong, S.S.H.; Lim, S.W.H. Mental imagery boosts music compositional creativity. *PLoS ONE* **2017**, *12*, e0174009. [CrossRef] [PubMed]
37. Commodari, E.; Sole, J.; Guarnera, M.; La Rosa, V.L. Mental imagery in education: What impact on the relationships with visuospatial processing and school performance in junior high school students? *Think. Ski. Creat.* **2024**, *54*, 101667. [CrossRef]
38. Schutte, N.S.; Malouff, J.M. A meta-analysis of the relationship between curiosity and creativity. *J. Creat. Behav.* **2020**, *54*, 940–947. [CrossRef]
39. Ryan, R.M.; Deci, E.L. Self-determination theory. In *Encyclopedia of Quality of Life and Well-Being Research*; Springer International Publishing: Cham, Switzerland, 2024; pp. 6229–6235.
40. Op den Kamp, E.M.; Tims, M.; Bakker, A.B.; Demerouti, E. Creating a creative state of mind: Promoting creativity through proactive vitality management and mindfulness. *Appl. Psychol.* **2023**, *72*, 743–768. [CrossRef]
41. Huo, M.; Ning, B. Network analysis of social-emotional skills in children and adolescents across ten cities: Evidence from the 2019 OECD Survey on Social and Emotional Skills. *Soc. Sci. Med.* **2025**, *381*, 118286. [CrossRef]
42. Yao, H.; Chen, S.; Liu, A. Exploring the relationship between academic challenge stress and self-rated creativity of graduate students: Mediating effects and heterogeneity analysis of academic self-efficacy and resilience. *J. Intell.* **2023**, *11*, 176. [CrossRef]
43. Wang, L.; Cui, Y.; Wang, X.; Wang, J.; Du, K.; Luo, Z. Regulatory focus, motivation, and their relationship with creativity among adolescents. *Front. Psychol.* **2021**, *12*, 666071. [CrossRef]
44. Carvalho, M.; Branquinho, C.; Moraes, B.; Cerqueira, A.; Tomé, G.; Noronha, C.; Gaspar, T.; Rodrigues, N.; de Matos, M.G. Positive Youth Development, Mental Stress and Life Satisfaction in Middle School and High School Students in Portugal: Outcomes on Stress, Anxiety and Depression. *Children* **2024**, *11*, 681. [CrossRef] [PubMed]
45. Thomson, P. *Resilience and Adaptation*; Elsevier: Amsterdam, The Netherlands, 2020.
46. López-Aymes, G.; Acuña, S.R.; Ordaz Villegas, G. Resilience and creativity in teenagers with high intellectual abilities: A middle school enrichment experience in vulnerable contexts. *Sustainability* **2020**, *12*, 7670. [CrossRef]
47. Ramírez, Y.; Navas, M.; López, V. Un estudio sobre la creatividad, el género, la edad, y las inteligencias múltiples en alumnos de educación secundaria obligatoria en España. *Prax. Educ.* **2019**, *23*, 1–16. [CrossRef]
48. Solhi, M.; Salehfard, A.; Ganji, M. Relationship between self-regulated strategies and creativity with the academic performance of public health students. *Psychology* **2016**, *1*, 53–62.

https://doi.org/10.3390/children13020261

children

MDPI

Review

Inclusion Goals: What Sex Education for LGBTQIA+ Adolescents?

Stefano Eleuteri *, Marta Girardi, Rossella Spadola and Elisabetta Todaro

Institute of Clinical Sexology, 00198 Rome, Italy; martagirardi@hotmail.it (M.G.); roxyspa77@gmail.com (R.S.); elisabettatodaro@gmail.com (E.T.)
* Correspondence: stefano.eleuteri@uniroma1.it

Abstract: According to the World Health Organization and the World Association for Sexual Health (WAS) Declaration of Sexual Rights, sex education aims to provide children and adolescents with comprehensive knowledge, skills, attitudes, and values, promoting equality and nondiscrimination while upholding freedom of thought and expression. However, current school curricula often focus on dichotomous sexual identity and hetero-cis-normative prevention strategies, neglecting the needs of gender minority (GM) and sexual minority (SM) youths. In family settings, discussions typically revolve around sexual risk reduction and basic contraception, omitting relational aspects and components of sexual identity such as orientations and gender expressions. This discrepancy highlights a gap between the official goals of sex education and its practical implementation, reflecting a cultural deficit in familial sex education. This study reviews the scientific literature on sexual health promotion interventions from 2015 to 2024 to identify inclusive approaches that enhance the participation of all youths, not just hetero-cis-normative individuals. The hypothesis is that most interventions primarily serve heterosexual and cisgender youths, indicating a need for more inclusive strategies to achieve better sexual health and educational outcomes. The study also suggests expanding curricula to align with recognized guidelines and the diverse needs of youth.

Keywords: inclusive sex education; LGBTQIA+ youths; sexual rights; parent-adolescent sex communication; school programs; sex educator professional program

1. Introduction

According to the World Health Organization [1,2] and the World Association for Sexual Health (WAS) Declaration of Sexual Rights [3], sex education aims to provide children and adolescents with comprehensive knowledge, skills, values, and attitudes promoting equality and nondiscrimination while upholding freedom of thought and expression.

The WHO-Regionalbüro für Europa und BZgASalute Standards [4] state that sexuality education means "learning relative to the cognitive, emotional, social, relational, and physical aspects of sexuality" and enhancing "the empowerment of children and young people by providing them with positive information, skills, and values for understanding their own sexuality and enjoying it, having safe and rewarding relationships, and behaving responsibly with respect to their own and others' sexual health and well-being. (. . .)" (p. 17).

At an international level, governments and education policy makers are reviewing the best practices in the delivery of relationship and sexuality education (RSE) [5]. Even if there are some changes to the most recent guidelines, like the New Zealand ones [6], modifications are mostly connected to a greater recognition of gender diversity rather than the active inclusion of diverse sexualities. These other issues could be mitigated by reimagining sexuality education in a way that could include explicit reference to sexual practices but also the possibility to adopt a 'norm-critical' approach to sexuality education, like the use of innovative strategies such as the careful choice of language to avoid unnecessary gendering and the adoption of an overt 'norm-critical' approach to avoid assumptions [7].

In other jurisdictions, inclusivity is more take into consideration, but is focused on cultural diversity rather than LGBTQIA+ perspectives [8].

For several years now, what were previously distinguished as different branches of sex education ("Sex Education" and "Comprehensive Sex Education"), have converged into the single representation of holistic sex education, promoting an integrated (between informational and emotional aspects), sex-positive, and inclusive perspective, accessible according to an intersectional view.

Adolescence can be considered as one of the most critical transitions in the life of the individual, during which facing a process of growth and change, characterized by different levels of maturation (physical, sexual, psycho-emotional, etc.), pushes adolescents toward the acquisition of dimensions such as social independence, identity development, and relational and emotional competencies.

As ascertained by leading scholars in Developmental Psychology and Developmental Psychosexology, this process may be more complex in cases where the individual is faced with the development of a non-heterosexual or non-cisgender sexual identity [9]. In this scenario, an additional element of difficulty may arise precisely from being differently included in sexual health promotion programs [10].

Talking about non-heterosexual and non-cisgender identities refers to movements to deconstruct the hetero-cis-normative framework, which defines "natural" and "normal" identities and behavior based on a complementarity between male and female. According to the hetero-cis-normative view, gender is thought of as corresponding to biological sex and as a dichotomous variable, and male and female are found to be mutually attractive, for both romantic and sexual and sentimental purposes. In sexuality education programs, one of the main limitations represented by this view in the context of sex education is the rigid prescription of stereotypes regarding gender identity and gender expression, as well as limiting knowledge and exploration of all legitimate components of sexual identity. LGBTQIA+ is the abbreviation for "Lesbian, Gay, Bisexual, Transgender, Queer, Intersex, and Asexual" people. The additional "+" stands for all of the other identities not encompassed in the shorter acronym. It is an umbrella term that is often used to refer to the community of non-heterosexual and non-cisgender identities as a whole.

According to evidence from the scientific literature, sex education is more effective on heterosexual youth than on LGBTQIA+ youth [11]. This finding can be explained by the insufficient inclusion of Sexual Minorities (SMs) and Gender Minorities (GMs) within programs. Therefore, the current most widely used sex education is poorly able to meet needs, identities, and subjectivities, entailing the risk of being an information and growth tool only for a narrow part of the target audience for which it is designed. Thus, it may be interesting to have some risk-related data. In general, SM and GM youths experience more negative outcomes, relative to sexual health, than heterosexual peers; for example, homosexual male youths have an extremely higher risk of HIV infections [10].

As evident from Figure 1, SM youth who are recipients of hetero-cis-normative sex education, in addition to being exposed to greater sexual risks, also reveal worse psychological health status, reporting higher rates of attempted suicide, substance abuse, depression, and anxiety, compared to their heterosexual cisgender counterparts [12–15].

What Figure 1 intends to show is the multidimensional (sexual, psychological, and relational) complex of negative factors that hetero-cis-normative culture entails for the health of young people belonging to SMs and GMs. Indeed, in societies and cultures strongly anchored in the hetero-cis-normative framework, the absence of a constructivist view of sexual identity and the lack of promotion of free self-knowledge can lead young people to incur in nonconscious, unwanted, and non-respectful experiences.

Effects such as these can be discussed by correlating them with the stressors that, following the Minority Stress model [16], are specific to SMs and GMs, originating phenomena such as internalized homo/transphobia, stigma, bullying, discrimination, and violence [15].

LGBTQIA+ youths who face bullying and bias in educational settings are more likely to suffer academically, displaying lower grade point averages, higher rates of absenteeism, a sense of alienation from their school environment, and diminished self-esteem, compared to their peers who do not encounter such victimization and discrimination at school [17].

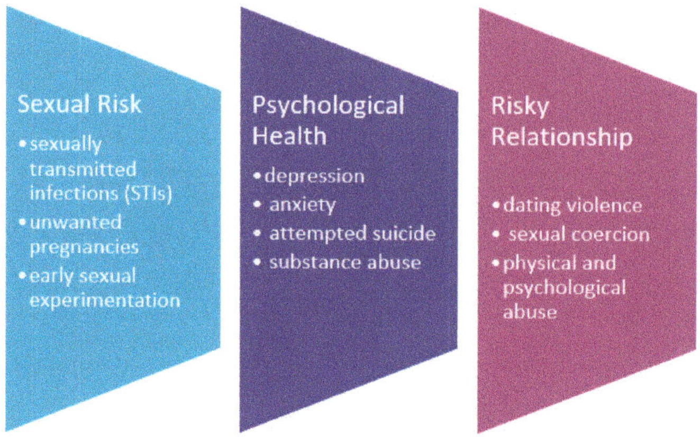

Figure 1. Effects of hetero-cis-normativity sex education.

The objective of this article is to review the published scientific literature in the field of sex education in order to highlight the presence or absence of inclusive programs and methodologies of work in the adolescent target audience; a further aim is to advance a critical discussion of the effects of the currently most popular programs and proposals for the use of more inclusive intervention methodologies [18,19]. The use of inclusive sex education aims to decrease school-based victimization that characterizes the experiences of SM and GM youth, the impact of sexually transmitted diseases (of which LGBTQIA+ youth are found to be one of the most at-risk groups), and the risk of experiencing violent relationships [20].

LGBTQIA+ students who received inclusive sexual health education showed the following outcomes: diminished levels of victimization, higher feelings of safety at school, less safety-related school absences, increased academic performance, and more feelings of connection with peers [21].

Evidence-based programs have also shown having an impact on sexual behavior. However, these programs are often not inclusive of LGBTQIA+ identities, thus limiting program effectiveness in the general population.

Programs that have taken more inclusive measures still have gaps, especially in the relationship aspect.

Recent research [22] has showed that digital spaces can be positively correlated with social support, self-esteem, diminished sense of loneliness, increasing identity socialization, expression of emotions, and development of positive interpersonal relationships.

It is important that practitioners and families consider digital tools as valuable allies, and turn their attention to a more inclusive and participatory promotion of sex education using both these tools and traditional ones.

In considering the use of a systemic perspective (which intends to involve all significant adults in the educational intervention on sexuality), it is important to consider educational action as a process of community empowerment and as a political action. Through educational processes on sexuality, it is possible to promote more inclusive and respectful communication skills, and to enhance dialogue within family units and peers. One example among all may be the recent debate on the use of inclusive language, which has raised (in some countries such as Italy) trends of opposition to the possibility of dealing with the topic of gender identity within school sex education curricula, especially in elementary school. At the same time, it has raised the debate on the topic, leading many parents, teachers, and educators to confront and activate processes of questioning some of the stereotypes present in the culture of reference.

2. Materials and Methods

This narrative review is based on the results produced by a search of several databases, such as Academic Search Index, CINAHL, Complementary Index, EBCO, Scopus, Medline, MLA International Bibliography, PubMed, Science Direct, and Springer Nature. The selected articles refer to publications dated from 2015 through 2024. The keywords used for the search are "LGBTQIA+ comics", "LGBTQIA+ inclusive sex education", "LGBTQIA+ narratives", "LGBTQIA+ youth", "sexual communication", "LGBTQIA+ sexual education", and "LGBTQIA+ students" AND "parent-adolescent sex communication". Based on the results obtained and the purpose of this paper, the most relevant studies were selected. Subsequently, the references and citations in the articles were used to expand the documentation. The data from the studies and the authors' positions were analyzed to frame the current "state of the art" in order to identify critical issues in current inclusive programs and promote new educational intervention strategies. To be included in this review, a quantitative study had to meet all the following inclusion criteria:

a. should have been published in peer-reviewed journals during or since 2015;
b. should have focused on LGBTQIA+ population;
c. should have been published in English and Spanish language.

Papers were excluded if:

a. inclusive criteria were not present.

Since this is a narrative review, no other guidelines were followed during the search and the selection of the studies.

In Figure 2, a flow chart of the study selection procedure is provided.

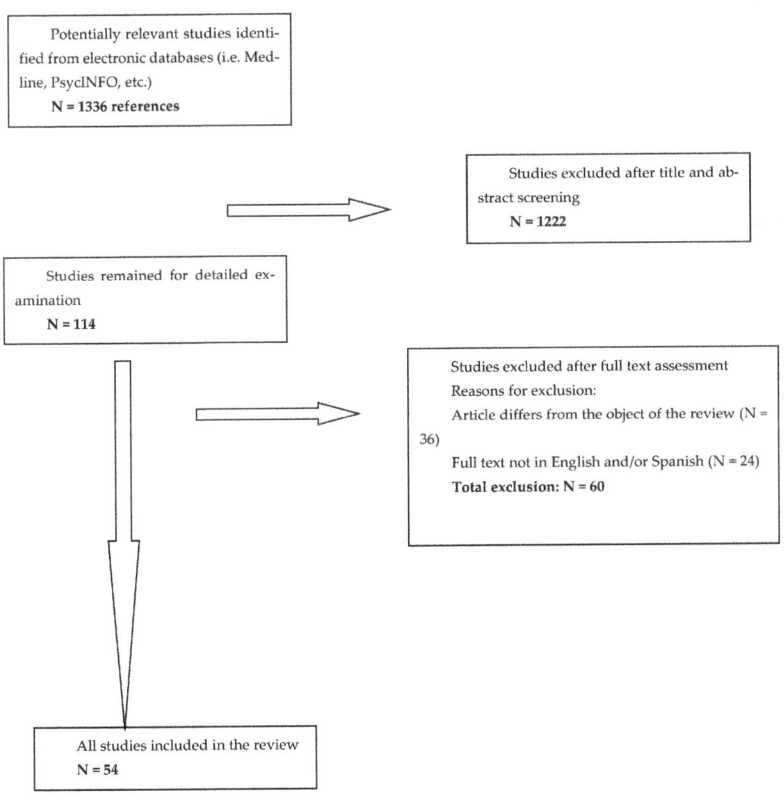

Figure 2. Flow chart of study selection procedure.

3. Results

3.1. The Effects of Sex Education Targeting Heterosexuals

Most of the evidence on inclusive policies comes from the United States; it is generally focused on the impact of school-level policies rather than state-level policies [23,24]. In a study conducted by Evans, heterosexual young adult women reported receiving more information about sex before becoming sexually active than bisexual or lesbian young adult women [25]. Another research suggests that transgender identities are completely excluded from the contents addressed in sexuality education [26]. Regarding sexual risk, overall, SM and GM youths experience more negative sexual health outcomes than their heterosexual peers. It is found in the literature that gay and bisexual women run a higher risk of sexually transmitted infections (STIs) and unwanted pregnancies [27]. In addition, among transgender youth, those younger than 13 years old are three times more prone to experience sexual experimentation [28].

3.2. Suggestions for Improving Inclusive Sex Education

Research highlights the importance of dedicated youth work for LGBTQ+ youth, providing them with a safe, nonjudgmental, and welcoming space to explore their identities with professionals who can serve as positive LGBTQIA+ role models and to have the opportunity to network with peers what they can share the journey of knowing their identities [29,30].

Current sex education programs have taken measures to be more inclusive, but they still have gaps, especially in the relationship aspect, and students rarely actually receive sex health education inclusive of LGBTQIA+ population [31]. In the educational curriculum reported in the Chicago City Agency Newsletter, starting in 2nd grade (7–8 years old), students learn the distinction between gender and sex. In the 5th grade (ages 10–11), the specific topic of LGBTQIA+ identities is introduced, which is then brought up again in the 13–16 age range. In addition, issues such as the risk of STIs are addressed through various types of sexual stimulation (such as oral, anal, manual, and vaginal intercourse), and issues like insemination, adoption and surrogacy are considered for heterosexual and LGBTQIA+ families. Finally, guidelines have been adopted to support transgender and gender-nonconforming youths by ensuring they have access to facilities that legitimize them and allow them to affirm their gender identity in all school settings [32].

3.3. Perceptions of the Youth Population

In 2018, Hobaica and Know studied SM and GM students' perceptions of sexuality education programs, finding that participants perceived their sexuality education to be permeated with normative elements.

Absence of visibility of all different sexual orientations, shame, and a lack of information on safe sexual practices emerged. This results in a feeling of unpreparedness for sexuality, which leads LGBTQIA+ youth to use other resources for self-education [33].

As summarized in Figure 3, a study of the sex and relationship education (SRE) experiences of young bisexual and queer women in Tasmania [34] found that most had received information on menstruation, puberty, contraceptive methods, information on preventing STIs, and reproduction. However, only a small percentage reported addressing issues related to gender, sexuality, and relationships. Complaints from participants in SRE programs included an excessive focus on biological aspects and a reproduction of gender inequalities through a focus on risk, personal responsibility, ethics, and punitive self-regulation, as well as an exclusive focus on risk in heterosexual relationships [34]. The points highlighted by this study, as shown in Figure 3, need to be taken into account when rethinking sex education programs from a perspective that can cross-cut the interests of all youth targets; the question educators can ask is: Is it possible to make the bio-psycho-social elements more inclusive? An example may be sexual anatomy: is there a way to be able to represent everyone's experience and perception?

Figure 3. Focus and limits of sex and relationship education.

An additional study conducted in the United States [35] collected information on the primary and secondary school sex education experiences of transgender youth (AMAB, AFAB, nonbinary) across a wide range of programs. These reports severely lack information on various identities, such as the complete absence of information on queer identities.

The experience of invisibility that characterizes the experiences of young people belonging to SMs and GMs with respect to sexuality education means that it is perceived as "uncomfortable" and not very conducive to the possibility of formulating the questions and elaborating on them in the growth of identity awareness and understanding. Pregnancy, abortion, STIs, and slut-shaming are the topics mainly addressed to promote fear and abstinence-oriented behaviors [36]. In addition, anatomical and biological aspects are often addressed in sex education programs with a division of classes according to a binary gender criterion (males and females) resulting in the exclusion of individuals belonging to GMs. Complaints from participants about sexual education programs concerned an excessive focus on biological aspects and a reproduction of gender inequalities through a focus on risk, personal responsibility, and self-regulation that was moralizing and punitive, as well as focused exclusively on risks in heterosexual relationships [37].

Regarding the period of puberty, transgender participants reported experiencing it with feelings of distress and alarm, particularly at the onset of gender dysphoria and the subsequent concerns associated with body image.

Figure 4 summarizes the main areas of intervention that should be addressed to achieve inclusive sex education. In continuity with what has been proposed in the bio-psycho-social approach, a sex education able to achieve the goal of developing and enhancing a positive attitude to the body and pleasure, as well as a healthy and respectful relationship with self and others, needs to work on several areas, such as: assertiveness (respect for one's rights and sexual safety), body image, the expression of one's sexual identity, and knowledge, free from taboos and prejudices.

Indeed, due to a description of body changes that did not propose representations accessible to transgender people, a perception was generated in them that their bodies would inevitably undergo changes, a source of inevitable discomfort [35]. In some parts of the world, so-called "homosexual acts" are prohibited by law resulting in prison sentences and, in some cases, the death penalty. The existence of laws such as these imply on LGBTQIA+ people, regardless of their narrow sphere of daily life, resistance and discomfort in expressing their sexual identity (i.e., the phenomenon called "in the closet"). In younger people, this state can lead to delays and discomfort in knowing, being aware of, and exploring one's gender, sexuality, and relational experience [38]. People in the LGBTQIA+ community may, in fact, find themselves struggling to learn what it means to form a relationship and having to defend themselves from family members who do not accept their identity [39].

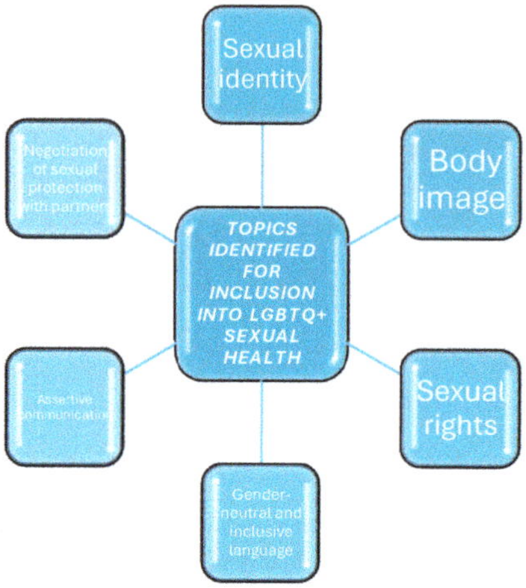

Figure 4. Perspectives and educational needs of LGBTQ+ youth.

3.4. *The Importance of the Web Context*

In this landscape, LGBTQIA+ people are trying to fill the void of institutionalized sex education by sharing important information about sexual health and rights. Today, social media is a new dissemination medium used by SMs and GMs to achieve this goal [40].

For LGBTQIA+ youths, online access to information gives the opportunity for an equality of access. Online sexual health information has been shown to provide LGBTQIA+ youths with control over their sexual health needs, giving them the opportunity to self-educate on topics not covered in traditional sexuality education, like diverse sexuality, interventions (surgical and non) for transgender individuals, and gender identities [41].

3.5. *The Importance of Inclusive Dialogue with Key Adults: The Role of Parents, Teachers, and Educators*

It seems that promotion of sexual knowledge by the family can be effective in improving sexual safety. It seems, moreover, that family-based educational interventions, in which parent–child communication skills are improved, lead to significant reductions in the enactment of risky sexual behaviors [25].

Parent–child LGBTQIA+ communication about sexuality can be fraught with barriers, as youth often experience anxiety, fear, and discomfort. Young people often avoid initiating conversations about sex after hearing negative comments from their parents about SM and GM people. Moreover, in some cases, parents are believed to lack useful information. Parental support is fundamental in influencing youths' willingness to engage in conversations regarding sex, making them more receptive and proactive to issues pertaining to sexual health promotion.

Topics such as romantic relationships, emotions, and dating safety may be rare in communication between parents and SM and GM children [42].

An interesting part of studies also focuses on understanding aspects supporting environmental and structural elements as a form of inclusion in schools.

Beasy and collaborators examined teachers' and staff members' perceptions, underscoring how the design and arrangement of spaces can either facilitate or hinder inclusion [43]. "Safe spaces" are particularly significant; they are environments where LGBTQIA+ students can express their identities without fear of discrimination or bullying. The study suggests that creating these spaces involves not only physical considerations—such as designated areas for LGBTQIA+ discussions and activities—but also the establishment of supportive relationships among staff and students. This can include training for teachers on LGBTQIA+ issues, promoting awareness and acceptance among the broader student body, and actively engaging LGBTQIA+ students in the process of defining what a safe space means to them.

Furthermore, the research points to the importance of ongoing dialogue among educators about their roles in creating inclusive environments. By negotiating and establishing these safe spaces, teachers and staff can contribute to a more supportive school culture that values diversity and fosters a sense of belonging for all students.

This approach aligns with broader educational goals of equity and inclusion, demonstrating how environmental and structural elements can be leveraged to support marginalized groups within educational settings.

In school settings, professionals involved in adolescent sex education recognize that this environment can be dangerous for SM and GM youth as victims of discrimination [44].

GLSEN (Gay, Lesbian, and Straight Education Network), a U.S.-based educational organization that addresses bullying, harassment, and discrimination based on gender identity, gender expression, and sexual orientation, has found that non-binary students perform worse academically than heterosexual students, but also compared to their lesbian, gay, and bisexual counterparts. According to data reported (2015), feelings of unsafety, fear, and experiences of harassment related to gender expression, discomfort, and gender-related insults are the motivations that lead young people to avoid and/or negative experiences with respect to the school environment.

A key finding to consider is that transphobic comments did not come from peers but, for 63.5% of respondents, from teachers and/or other adult figures in the school setting [44].

Among the adults involved, educators play an important role in this area. Therefore, it is a striking finding that most education professionals indicate that they do not have adequate knowledge to work with SM and GM students [45].

In the meta-analysis on LGBTQ-inclusive sexuality education, O'Farrell et al. (2021) discuss challenges faced by teachers. They suggest that there is an ambivalence and anxiety from some facilitators to carry on inclusive sexuality education that can be explained with their own stigma or perceived inability to talk about these topics because of a lack of information and training.

The literature, in fact, states that skills related to affirming young people's identities, using neutral and specific language and creating relevance, are not often considered in teachers' training [41].

Educators require training that broadens their understanding of biological sex, gender identity, sexual and romantic orientation, and sexual behavior.

The recognition that language is constantly evolving and that young people have the right to self-identify should be integrated into the teaching of basic terminology.

Considering that the biopsychosocial approach appears to be the most comprehensive in the implementation of effective sex education interventions, it is believed that the use of the integrated approach (which works on cognitive, emotional, and socio-cultural aspects regarding sexuality) can be a good reference for the training of educators. Through the integrated training methodology, it is possible to propose the study of the most effective models, in parallel with the experimentation of training experiences (through simulations, bioenergetics exercises, and graphic and creative exercises) that allow future sexuality educators to become more aware of implicit and explicit messages, taboos, personal beliefs, and values that they carry in their interventions. In this way, it is possible to imagine that educators can convey messages and content that are more inclusive and able to increase the self-determination resources of the target youth.

A recent study conducted in Europe [46] confirmed the importance of inclusive school policies for LGBTIyouth. Key takeaways include:

- Improved Safety and Well-Being: Schools with more inclusive policies lead to lower odds of lack of safety and concealment among LGBTQIA+ youth. This indicates that when schools actively promote inclusivity, students feel safer and more accepted.
- Higher Life Satisfaction: LGBTQIA+ youth in inclusive environments report higher levels of life satisfaction, suggesting that acceptance and support within the school context significantly enhance their overall well-being.
- Impact of Teacher Training: Training teachers to be more inclusive positively affects school climate. It is associated with reduced feelings of depression and sadness among LGBTQIA+ youth, as well as reduced bias-based violence.
- Role of Inclusive Curricula: Including LGBTQIA+ topics in the curriculum not only promotes visibility but also correlates with decreased experiences of harassment, both general and bias-based. This highlights the importance of representation in educational materials.
- Visibility vs. Concealment: Teacher training is linked to greater visibility of LGBTQIA+ youth, which in turn reduces their need to conceal their identities. This suggests that supportive educators can create an environment where students feel comfortable being themselves.

Overall, the study emphasizes that comprehensive strategies involving both teacher training and inclusive curricula are essential for fostering a safe and supportive school environment for LGBTQIA+ youth. By equipping educators with the necessary skills and knowledge, and integrating inclusive content into the curriculum, schools can create an environment where LGBTQIA+ students feel accepted and protected. This dual approach addresses both the immediate need for a supportive atmosphere and the long-term goal of systemic change in attitudes and practices within the educational system.

Moreover, the findings underscore the critical role that inclusivity in teacher training and curricula plays in fostering a safer school environment for LGBTQIA+ youth. By incorporating comprehensive training that emphasizes diversity and the enumeration of protected groups, teachers are better equipped to create a supportive atmosphere that actively reduces feelings of depression and sadness among these students (by 10%). Moreover, these inclusive practices are essential in combating bias-based school violence, highlighting the need for educational institutions to prioritize training that promotes understanding and acceptance.

Increased inclusivity in teacher training and curricula was significantly linked to reduced likelihood of perceived lack of school safety (by 17% and 16%, respectively) and to 18% and 14% lower odds of bias-based school violence, respectively.

The significant reduction in the odds of perceived school safety issues and bias-based violence suggests that when educators are trained to recognize and value diversity, they can effectively model positive behavior and set norms that discourage bullying [47]. This aligns with existing literature that emphasizes the necessity of specialized skills in addressing bias-based victimization, indicating that targeted interventions can lead to substantial improvements in school climate and student well-being.

Overall, these findings advocate for the implementation of inclusive policies and training programs as a means to enhance safety and support for marginalized youth, ultimately contributing to a more equitable educational environment. Often, bias-based victimization prevention efforts require specific knowledge and skills [48].

A recent study aimed to gain a comprehensive understanding of the LGBTQIA+ community's presence in sexuality education and the ways it is addressed. To achieve this, qualitative research was conducted using a collective case study approach within the context of sexuality education activities at two secondary schools in the province of Almería (Andalusia, Spain). Data collection methods included non-participatory observation, semi-structured interviews, and analysis of documentary sources. The findings indicate a significant prevalence of exclusionary perspectives, the perpetuation of discriminatory beliefs, and a tendency to silence. Furthermore, they demonstrate that the LGBTQIA+ community is treated in a superficial manner. Consequently, sexuality education is not viewed as a space for discussing non-hegemonic identities and experiences, nor are these realities recognized as part of the broader spectrum of sexuality. This perspective contributes to the continuation and legitimization of a stigmatized viewpoint, as non-hegemonic identities and experiences are seen as abnormal, rare, exceptional, and taboo [49].

4. Discussion

4.1. Influence of SRE in LGBTQIA+ Youths

One of the most important factors that has emerged as influencing SRE in GM and SM youths seems to be their relationship with parents. In fact, communication between parents and LGBTQIA+ adolescents may be characterized, prior to their child's coming out, by a hetero-cis-normative approach [50]. In this case, parents may have more difficulty, as they may be less aware, educated, and comfortable about same-sex sexual behaviors, or raising a transgender or nonbinary person, as well as being influenced by general and, potentially, negative beliefs and attitudes.

In Australia [51], research has found that LGBTQIA+ youth, unlike their heterosexual counterparts, perceive SRE programs as the least useful source of information. In contrast, they regard online information and social media as the most important sources. This datum seems to be strictly connected to the themes that are discussed in such interventions. Regarding the themes, it emerged from a study that from ages 14 to 15, certain issues receive more consideration, like the use of gender-neutral and inclusive language, and the utilization of gender-neutral terms like "partner" rather than "boyfriend" [52–54].

4.2. Use of a Neutral and Specific Language to Create an Inclusive Environment

Neutral language and specificity both play important roles in creating inclusive environments, particularly in educational settings.

Neutral language serves as an umbrella that encompasses a diverse range of identities, allowing individuals from various sexual orientations and gender identities to feel represented and included. This approach can foster a sense of belonging among young people who might otherwise feel marginalized or overlooked in traditional curricula. By using language that is inclusive and non-specific, educators can create a welcoming atmosphere that encourages all students to engage with the material without fear of exclusion.

On the other hand, specificity acknowledges and names individual identities, providing visibility to particular groups that may otherwise remain invisible in neutral language. This can help students see themselves in the scenarios presented, facilitating deeper reflection on their own experiences and the decisions they face [15]. When specific identities are highlighted, it can lead to richer discussions and a better understanding of the unique challenges and perspectives that different individuals encounter.

Both approaches have their merits: neutral language promotes inclusivity and broad appeal, while specificity fosters recognition and validation of individual experiences. Striking a balance between the two can enhance educational practices, ensuring that all students feel seen, heard, and valued.

4.3. What Topics to Include for Inclusive Sex Education?

A 2014 study by the U.S. Centers for Disease Control and Prevention (CDC) [55] found that 72% of high schools in the United States educate students on how to prevent pregnancy, 60% promote the effectiveness of contraception, and 35% instruct students in condom use.

It would be important for them to be included, in order to address issues regarding consent, safety, assertive communication, and negotiating sexual protections with partners. These aspects are of paramount importance, even more so when considering that young people belonging to SMs and GMs are found to be more at risk for sexual coercion, dating violence, and physical and psychological abuse [56].

Comprehensive sex education that includes discussions on consent, safety, and communication can provide them with the tools to navigate these challenges more effectively.

Teaching students about consent is crucial in fostering a culture of respect and mutual agreement in sexual activities. Ensuring safety ensures that students are better prepared to protect themselves in various contexts. In the context of sexual health, teaching students how to communicate assertively helps them to develop healthy relationship dynamics where both partners can voice their opinions and concerns without fear of judgment or retribution.

This education is essential in creating a safe environment where individuals feel empowered to make informed choices about their bodies and relationships.

The CDC's findings highlight the progress made in sexual health education in U.S. high schools, but also underscore the gaps that still exist.

4.4. Narrative Approach in Educational Programs for an Inclusive Sexuality

Building on Plummer's (1995) [57] work focused on sexual histories, the importance of adopting a "narrative" perspective in interventions on identity components has been increasingly reinforced in order to capture the dynamic meaning of the dimensions that structure human sexuality. Adopting a narrative perspective in interventions on identity components is crucial for capturing the dynamic and multifaceted nature of human sexuality. This approach allows for a more comprehensive and empathetic understanding of sexual identities, fostering a more inclusive and supportive environment for individuals to explore and express their sexuality. This approach contrasts with static models of identity, which often fail to capture the lived realities of individuals. Plummer's work highlights the importance of these personal stories in understanding the diversity and fluidity of human sexuality.

Gilbert [58] argues that in order to recognize the rights of people in the LGBTQIA+ community, it is necessary to adopt a theory of sexuality that both tolerates and, at the same time, challenges the relationship one has to the categories used both from outside and within the community [59]. By focusing on the narrative aspects of sexual histories, interventions can also challenge and deconstruct harmful societal norms and stereotypes. This approach encourages a more inclusive understanding of sexuality, recognizing the validity of diverse sexual identities and experiences. It aligns with contemporary views on sexuality, which emphasize the importance of personal agency and the right to define one's sexual identity free from societal constraints.

The issue of identity has been recognized [32] as a central aspect to be considered in sexuality education programs, including a welcoming and representative view of the different forms it can take, overcoming the conception of a "normative" and standardized pathway. In addition to the importance they represent for people living with commitment to the delicate topic of identity exploration, issues such as coming out, the transition process, asexuality, and nonbinary identities prove to be crucial across the board.

In this regard, it may be significant to reflect on the fact that "inclusion" means "welcoming within"; the perspective that educators need to implement most is to offer reflections, analysis, debate, and critical discussion for all and on all aspects of knowing, exploring, and expressing one's sexual identity.

It is fundamental to consider that an individual may also belong to multiple SMs or GMs and/or other social groups and categories. Organizing interventions by grouping them into a single category, such as "sexual and gender minority", can be misleading and ineffective in promoting sexual health [60].

4.5. Use of Alternative Sources of Information

Among the tools that can be used in order to tell and see diverse experiences represented, two can be proposed: narratives and comics. Narratives are an opportunity to highlight the significance of diversity and contextual sensitivity when one's identity and Minority Stress are at stake, in situations where it is health that is being undermined: "Stories work well in representing the need, the difficulty and often the anger of choosing how to live" [61]. Through storytelling, people can articulate their experiences, desires, and challenges, providing a more nuanced view of their sexuality.

As far as LGBTQIA+ adolescents are concerned, an interesting option as an opportunity to read narratives is represented by comic books. Numerous lesbian, bisexual, and queer cartoonists have used this form of expression to popularize the various narratives aimed at this age group of the LGBTQIA+ population. Such representations provide an opportunity to examine how these stories are told, the use of iconographic details aimed at portraying romantic relationships between characters, and how representations of non-hetero-cis-normative relationships provide a positive view of LGBTQIA+ characters [62]. Seeing characters who share similar struggles, triumphs, and everyday experiences can be incredibly validating. This representation helps combat feelings of isolation and fosters a sense of belonging and visibility [63].

Moreover, the combination of visual art and storytelling in comics makes them particularly accessible and engaging. For adolescents, the visual aspect can make complex themes more understandable and relatable. The format often allows for a more immediate emotional connection, as facial expressions, body language, and visual metaphors complement the narrative [64].

The comic book medium encompasses a wide range of genres, from superhero tales to slice-of-life stories, fantasy, science fiction, and romance. This diversity means that there are narratives that can cater to various tastes and interests, all while providing meaningful LGBTQIA+ representation.

Delivering sexuality education focused on traditional heterosexuality (i.e., vaginal intercourse and reproduction) is not congruent with the daily experiences of youths today, many of whom are sexually fluid (and/or engage in a wide range of sexual practices regardless of sexual identity [54,65].

5. Conclusions

In light of the discussed contributions, it is of paramount importance that sex educators, but also all those who provide training to adolescents (teachers, parents, sport workers, and health professionals) engage in fully implementing the fundamentals of holistic sex education, which, as mentioned, involves an inclusive and cross-cutting epistemology [66–77].

Inclusive education should discuss of all different experiences and practices, offering a non-judgmental environment and the possibility of feeling included in the education. In order to do this, it is important to have both a formal and an informal sexual education, suggesting also possible useful websites and/or apps that could be useful [3,78]. Narratives and comics could be two important tools to use with this specific target [54,63].

Different studies have demonstrated diminished well-being for lesbian, gay, bisexual, transgender, queer, intersex, and asexual (LGBTQIA+) youths relative to their peers, with the most results focused on negative experiences at school as key factors associated with these differences [79–82].

All of this is central not only to make interventions more effective and more responsive to the needs of target audiences, but also to embrace the call of the technical documents to make sex education a process of growth, respect, and acceptance of all and for all.

In this regard, it is deemed necessary to further point out that knowing the specific educational needs of the LGBTQIA+ adolescent population does not mean working exclusively in the creation of ad hoc interventions.

Sexuality education that does not explicitly address diversity is ineffective for young people growing up in a world where sexual and gender diversity is prevalent and understandings of sexuality and gender differ significantly from previous generations. It is important that all young people—regardless of their gender identity—have access to appropriate and accurate information about sexuality that goes beyond the dominant heterosexist approach to include sexual and gender diversity. An inclusive approach to sexuality education should, at a minimum, mean that young people are well prepared for future consensual sexual relationships, whatever that may entail. Through an inclusive approach, all young people will have a better understanding of diversity.

The first limitation is that this is a narrative review, so a systematic one could provide data that are more detailed. Another limitation of the review is that it is based upon a compilation of peer-reviewed English and Spanish language literature. This literature largely comes from Europe and North America, much less often from Latin America and Australia, and Asian or African sources are completely absent. The studies referenced thus approach the subject of investigation from a predominantly Western perspective. The review provides little information on both the online sexual activities in non-Western populations and the non-English academic discourses on these activities. The last limitation that we think it is useful to address is that maybe different reviews could specifically concentrate on the different sexual and gender identities, in order to provide more detailed info on each of them.

What is proposed, instead, as a further perspective of development, is to broaden the attention and sensitivity in writing projects, defining themes, selecting tools and activities-stimulus, in order to make whole groups work, with all the different subjectivities and identities that compose and structure them. As highlighted by others [83,84] inclusivity in the wider school environment is central to LGBTQIA+ students' sense of safety and belongingness in the school context.

Author Contributions: S.E.: Conceptualization; Validation; Writing—Review and Editing. M.G.: Conceptualization; Data Curation; Writing—Original Draft; R.S.: Methodology; Validation; Writing—Review and Editing. E.T.: Conceptualization; Validation; Writing—Review and Editing. All authors have read and agreed to the published version of the manuscript.

Funding: This research received no external funding.

Conflicts of Interest: The authors declare no conflicts of interest.

References

1. World Health Organization. Sexual Health and Its Linkages to Reproductive Health: An Operational Approach. Available online: https://www.who.int/publications/i/item/978924151288 (accessed on 11 July 2024).
2. World Health Organization. WHO Recommendations on Adolescent Sexual and Reproductive Health and Rights. Available online: https://iris.who.int/bitstream/handle/10665/275374/9789241514606-eng.pdf (accessed on 12 July 2024).
3. World Association for Sexual Health. WAS Declaration: Sexual Health for the Millenium. Available online: https://www.worldsexualhealth.net/_files/ugd/793f03_779f193815894895b9d1480bde676c37.pdf?index=true (accessed on 12 July 2024).
4. WHO-Regionalbüro für Europa und BZgASalute. Standards für die Sexualaufklärung in Europa. Köln, Bundeszentrale für Gesundheitliche Aufklärung (BZgA). Available online: https://www.bzga-whocc.de/publikationen/standards-fuer-sexualaufklaerung/ (accessed on 11 July 2024).
5. Garg, N.; Volerman, A. A national analysis of state policies on lesbian, gay, bisexual, transgender, and questioning/queer inclusive sex education. *J. Sch. Health* **2021**, *91*, 164–175. [CrossRef] [PubMed]
6. Ministry of Education. Relationships and Sexuality Education: A Guide for Teachers, Leaders, and Boards of Trustees (Years 9–13). Available online: https://www.hauoratairawhiti.org.nz/assets/Uploads/RSE+Guide+y1-8.pdf (accessed on 12 July 2024).
7. Bengtsson, J.; Bolanderm, E. Strategies for Inclusion and Equality 'norm-critical' Sex Education in Sweden. *Sex Educ.* **2020**, *20*, 154–169. [CrossRef]
8. Krebbekx, W. Including Diversity? The Politics of Sex Education in The Netherlands. In *SAGE Handbook of Global Sexualities*; Davy, Z., Santos, A.C., Bertone, C., Thoreson, R., Eds.; SAGE: London, UK, 2020; pp. 207–220, ISBN 9781529721683.
9. Flores, D.; Docherty, S.L.; Relf, M.V.; McKinney, R.E.; Barroso, J.V. "It's Almost Like Gay Sex Doesn't Exist": Parent-Child Sex Communication According to Gay, Bisexual, and Queer Male Adolescents. *J. Adolesc. Res.* **2019**, *34*, 528–562. [CrossRef] [PubMed]
10. Charley, C.; Tureson, A.; Wildenauer, L.; Mark, K. Sex Education for LGBTQ+ Adolescents. *Curr. Sex. Health Rep.* **2023**, *15*, 180–186. [CrossRef]
11. Bodnar, K.; Tornello, S.L. Does sex education help everyone?: Sex education exposure and timing as predictors of sexual health among lesbian, bisexual, and heterosexual young women. *J. Educ. Psychol. Consult.* **2019**, *29*, 8–26. [CrossRef]
12. Epps, B.; Markowski, M.; Cleaver, K. A rapid review and narrative synthesis of the consequences of non-inclusive sex education in UK schools on lesbian, gay, bisexual, transgender and questioning young people. *J. Sch. Nurs.* **2021**, *39*, 87–97. [CrossRef]

13. Heer, B.; Brown, M.; Cheney, J. Sexual consent and communication among the sexual minoritized: The role of heteronormative sex education, trauma, and dual identities. *Fem. Criminol.* **2021**, *16*, 701–721. [CrossRef]
14. Kaczkowski, W.; Cooper, A.C.; Li, J.; Robin, L. The association of LGBTQ-supportive school health policies and practices with sexual health outcomes. *LGBT Health* **2022**, *29*, 384–392. [CrossRef]
15. Kesler, K.; Gerber, A.; Laris, B.; Anderson, P.; Baumler, E.; Coyle, K. High School FLASH Sexual Health Education Curriculum: LGBTQ Inclusivity Strategies Reduce Homophobia and Transphobia. *Prev. Sci.* **2023**, *24*, 272–282. [CrossRef] [PubMed]
16. Meyer, I.H. Minority stress and mental health in gay men. *J. Health Soc. Behav.* **1995**, *36*, 38–56. [CrossRef]
17. Kosciw, J.G.; Greytak, E.A.; Zongrone, A.D.; Clark, C.M.; Truong, N.L. The 2017 National School Climate Survey: The Experiences of Lesbian, Gay, Bisexual, Transgender, and Queer Youth in Our Nation's Schools. Available online: https://www.glsen.org/sites/default/files/2019-10/GLSEN-2017-National-School-Climate-Survey-NSCS-Full-Report.pdf (accessed on 10 July 2024).
18. Naser, S.C.; Clonan-Roy, K.; Fuller, K.A.; Goncy, E.A.; Wolf, N. Exploring the experiences and responses of LGBTQ+ adolescents to school based sexual education. *Psychol. Sch.* **2022**, *59*, 34–50. [CrossRef]
19. Tabaac, A.R.; Johns, M.M.; Zubizarreta, D.; Haneuse, S.; Tan, A.S.L.; Austin, S.B.; Potter, J.; Lindberg, L.; Charlton, B.M. Associations between sexual orientation, sex education curriculum, and exposure to affirming/disaffirming LGB content in two US-based cohorts of adolescents. *Sex Educ.* **2022**, *23*, 506–523. [CrossRef] [PubMed]
20. Goldfarb, E.S.; Lieberman, L.D. Three decades of research: The case for comprehensive sex education. *J. Adolesc. Health* **2021**, *68*, 13–27. [CrossRef] [PubMed]
21. Day, J.K.; Fish, J.N.; Grossman, A.H.; Russell, S.T. Gaystraight alliances, inclusive policy, and school climate. *J. Res. Adolesc.* **2020**, *30*, 418–430. [CrossRef] [PubMed]
22. Mitchell, K.J.; Ybarra, M.L.; Korchmaros, J.D.; Kosciw, J.G. Accessing sexual health information online: Use, motivations and consequences for youth with different sexual orientations. *Health Educ. Res.* **2014**, *29*, 147–157. [CrossRef] [PubMed]
23. Russell, S.T.; Bishop, M.D.; Saba, V.C.; James, I.; Ioverno, S. Promoting school safety for LGBTQ and all students. *Policy Insights Behav. Brain Sci.* **2021**, *8*, 160–166. [CrossRef] [PubMed]
24. Ioverno, S.; Russell, S.T. Homophobic bullying in positive and negative school climates: The moderating role of gender sexuality alliances. *J. Youth Adolesc.* **2021**, *50*, 353–366. [CrossRef] [PubMed]
25. Evans, R.; Widman, L.; Goldey, K. The Role of Adolescent Sex Education in Sexual Satisfaction among LGB+ and Heterosexual Young Adults. *Am. J. Sex. Educ.* **2020**, *15*, 310–335. [CrossRef]
26. Andújar, A.G. Exclusiones y discriminaciones hacia las identidades trans en educación afectivo-sexual. *Aula Abierta* **2021**, *50*, 833–840. [CrossRef]
27. Doull, M.; Wolowic, J.; Saewyc, E.; Rosario, M.; Prescott, T.; Ybarra, M.L. Why girls choose not to use barriers to prevent sexually transmitted infection during female-to-female sex. *J. Adolesc. Health* **2018**, *62*, 411–416. [CrossRef]
28. Johns, M.M.; Lowry, R.; Andrzejewski, J.; Barrios, L.C.; Demissie, Z.; McManus, T.; Rasberry, C.R.; Robin, L.; Underwood, J.M. Transgender identity and experiences of violence victimization, substance use, suicide risk, and sexual risk behaviors among high school students—19 states and large urban school districts, 2017. *MMWR Morb. Mortal. Wkly. Rep.* **2019**, *68*, 67–71. [CrossRef]
29. Formby, E.; Donovan, C. Sex and relationships education for LGBT+ young people: Lessons from UK youth work. *Sexualities* **2020**, *23*, 1155–1178. [CrossRef]
30. McBride, R.S.; Neary, A. Trans and Gender Diverse Youth Resisting Cisnormativity in School. *Gend. Educ.* **2021**, *33*, 1090–1107. [CrossRef]
31. SIECUS (Sexuality Information and Education Council of the United States). SIECUS State Profiles 2020. Available online: https://siecus.org/state-profiles-2020/ (accessed on 11 July 2024).
32. Jarpe-Ratner, E. How can we make LGBTQ+-inclusive sex education programmes truly inclusive? A case study of Chicago Public Schools' policy and curriculum. *Sex Educ.* **2020**, *20*, 283–299. [CrossRef]
33. Keiser, G.H.; Kwon, P.; Hobaica, S. Sex Education Inclusivity and Sexual Minority Health: The Perceived Inclusivity of Sex Education Scale. *Am. J. Sex. Educ.* **2019**, *14*, 388–415. [CrossRef]
34. Grant, R.; Nash, M. Educating queer sexual citizens? A feminist exploration of bisexual and queer young women's sex education in Tasmania, Australia. *Sex Educ.* **2019**, *19*, 313–328. [CrossRef]
35. Hobaica, S.; Schofield, K.; Kwon, P. "Here's Your Anatomy... Good Luck": Transgender Individuals in Cisnormative Sex Education. *Am. J. Sex. Educ.* **2019**, *14*, 358–387. [CrossRef]
36. Hobaica, S.; Kwon, P. "This Is How You Hetero": Sexual Minorities in Heteronormative Sex Education. *Am. J. Sex. Educ.* **2018**, *12*, 423–450. [CrossRef]
37. McGlashan, H. Queer Youth and Schooling: A Critical Ethnography of How Queer Youth Experience and Navigate Their Intersectional and Fluid Identities at School. Ph.D. Thesis, University of Auckland, Auckland, New Zealand, 2021.
38. Mata, D.; Korpak, A.K.; Sorensen, B.L.; Dodge, B.; Mustanski, B.; Feinstein, B.A. A mixed methods study of sexuality education experiences and preferences among bisexual, pansexual, and queer (bi+) male youth. *Sex. Res. Soc. Policy* **2022**, *19*, 806–821. [CrossRef] [PubMed]
39. Schwartz, A.; Yarbrough, E.; McIntosh, C.A. Sex and Relationship Issues in Work with the LGBTQ Community. *Focus* **2020**, *18*, 277–284. [CrossRef]
40. Manduley, A.E.; Mertens, A.; Plante, I.; Sultana, A. The role of social media in sex education: Dispatches from queer, trans, and racialized communities. *Fem. Psychol.* **2018**, *28*, 152–170. [CrossRef]

41. O'Farrell, M.; Corcoran, P.; Davoren, M.P. Examining LGBTI+ inclusive sexual health education from the perspective of both youth and facilitators: A systematic review. *BMJ Open* **2021**, *11*, e047856. [CrossRef] [PubMed]
42. McKay, E.A.; Fontenot, H.B. Parent-Adolescent Sex Communication with Sexual and Gender Minority Youth: An Integrated Review. *J. Pediatr. Health Care* **2020**, *34*, e37–e48. [CrossRef] [PubMed]
43. Beasy, K.; Grant, R.; Emery, S. Multiple dimensions of safe space for LGBTQ students: School staff perceptions. *Sex Educ.* **2021**, *23*, 35–48. [CrossRef]
44. Kosciw, J.G.; Greytak, E.A.; Giga, N.M.; Villenas, C.; Danischewski, D.J. The 2015 National School Climate Survey: The Experiences of Lesbian, Gay, Bisexual, Transgender, and Queer Youth in Our Nation's Schools. Available online: https://files.eric.ed.gov/fulltext/ED590244.pdf (accessed on 9 July 2024).
45. Goodrich, K.; Barnard, J. Transgender and gender non-conforming students in schools: One school district's approach for creating safety and respect. *Sex Educ.* **2019**, *19*, 212–225. [CrossRef]
46. Ioverno, S. Inclusive National Educational Policies as Protective Factors for LGBTI Youth Adjustment: An European Cross-National Study. *J. Adolesc. Health* **2023**, *72*, 845–851. [CrossRef] [PubMed]
47. Ioverno, S.; Nappa, M.R.; Russell, S.T.; Baiocco, R. Student intervention against homophobic name-calling: The role of peers, teachers, and inclusive curricula. *J. Interpers. Violence* **2022**, *37*, NP19549–NP19575. [CrossRef] [PubMed]
48. Ioverno, S.; Bishop, M.D.; Russell, S.T. Does a Decade of School Administrator Support for Educator Training on Students' Sexual and Gender Identity Make a Difference for Students' Victimization and Perceptions of School Climate? *Prev. Sci.* **2022**, *23*, 108–118. [CrossRef]
49. Granero-Andújar, A.; Moreno-Morilla, C. A Collective Case Study in Spain about the LGBTI Community in Sexuality Education. *Multidiscip. J. Educ. Res.* **2024**, *14*, 202–222. [CrossRef]
50. Bouris, A.; Guilamo-Ramos, V.; Pickard, A.; Shiu, C.; Loosier, P.S.; Dittus, P.; Michael Waldmiller, J. A systematic review of parental influences on the health and well-being of lesbian, gay, and bisexual youth: Time for a new public health research and practice agenda. *J. Prim. Prev.* **2010**, *31*, 273–309. [CrossRef]
51. Robinson, K.H.; Bansel, P.; Denson, N.; Overden, G.; Davies, C. *Growing up Queer: Issues Facing Young Australians Who Are Gender Variant and Sexuality Diverse*; Young and Well Cooperative Research Centre: Melbourne, VIC, Australia, 2014; ISBN 9780987117984.
52. Pampati, S.; Johns, M.M.; Szucs, L.E.; Bishop, M.D.; Mallory, A.B.; Barrios, L.C.; Russell, S.T. Sexual and gender minority youth and sexual health education: A systematic mapping review of the literature. *J. Adolesc. Health* **2021**, *68*, 1040–1052. [CrossRef]
53. Narushima, M.; Wong, J.P.H.; Li, A.T.W.; Bhagat, D.; Bisignano, A.; Fung, K.P.L.; Poon, M.K.L. Youth perspectives on sexual health education: Voices from the YEP study in Toronto. *Can. J. Hum. Sex.* **2020**, *29*, 32–44. [CrossRef]
54. Ellis, S.J.; Bentham, R.M. Inclusion of LGBTIQ perspectives in school-based sexuality education in Aotearoa/New Zealand: An exploratory study. *Sex Educ.* **2020**, *21*, 708–722. [CrossRef]
55. Centers for Disease Control and Prevention. HIV and Young Men Who Have Sex with Men. Available online: https://www.cdc.gov/healthyyouth/sexualbehaviors/pdf/hiv_factsheet_ymsm.pdf (accessed on 10 July 2024).
56. Dank, M.; Lachman, P.; Zweig, J.M.; Yahner, J. Dating violence experiences of lesbian, gay, bisexual, and transgender youth. *J. Youth Adolesc.* **2014**, *43*, 846–857. [CrossRef] [PubMed]
57. Plummer, K. *Telling Sexual Stories: Power, Change and Social Worlds*; Routledge: London, UK, 1995; ISBN 9780203425268.
58. Gilbert, J. *Sexuality in School: The Limits of Education*; University of Minnesota Press Minneapolis: Minneapolis, MN, USA, 2014; ISBN 1452942226/9781452942223.
59. Clarke, K.; Cover, R.; Aggleton, P. Sex and ambivalence: LGBTQ youth negotiating sexual feelings, desires and attractions. *J. LGBT Youth* **2018**, *15*, 227–242. [CrossRef]
60. Steinke, J.; Root-Bowman, M.; Estabrook, S.; Levine, D.S.; Kantor, L.M. Meeting the Needs of Sexual and Gender Minority Youth: Formative Research on Potential Digital Health Interventions. *J. Adolesc. Health* **2017**, *60*, 541–548. [CrossRef] [PubMed]
61. Frank, A.W. *Letting Stories Breathe: A Socio-Narratology*; University of Chicago Press: Chicago, IL, USA, 2010; ISBN 0226260143/9780226260143.
62. Gillingham, E. Representations of same-sex relationships between female characters in all-ages comics: Princess Princess Ever After and Lumberjanes. *J. Lesbian Stud.* **2018**, *22*, 390–401. [CrossRef] [PubMed]
63. Faccini, B.; Di Giovanni, E. Diversity in Comics: LGBTQIA+ Characters and Their Importance in Adolescent Development. *J. Graph. Nov. Comics* **2021**, *12*, 77–92.
64. Gleason, K. Visual Storytelling and Emotional Engagement: A Study of Graphic Novels. *Comics Stud. J.* **2020**, *8*, 33–49.
65. Ellis, S.; Aitken, R. Sexual Health Practices of 16 to 19 Year Olds in New Zealand: An Exploratory Study. *J. Prim. Health Care* **2020**, *12*, 64–71. [CrossRef]
66. Atkinson, J.H.; Grant, I.; Kennedy, C.J.; Richman, D.D.; Spector, S.A.; McCutchan, J.A. Prevalence of psychiatric disorders among men infected with human immunodeficiency virus: A controlled study. *Arch. Gen. Psychiatry* **1988**, *45*, 859–864. [CrossRef]
67. Cochran, S.D.; Mays, V.M. Lifetime prevalence of suicide symptoms and affective disorders among men reporting same-sex sexual partners: Results from NHANES III. *Am. J. Public Health* **2000**, *90*, 573. [CrossRef] [PubMed]
68. Cochran, S.D.; Mays, V.M. Relation between psychiatric syndromes and behaviorally defined sexual orientation in a sample of the US population. *Am. J. Epidemiol.* **2000**, *151*, 516–523. [CrossRef]
69. Fergusson, D.M.; Horwood, L.J.; Beautrais, A.L. Is sexual orientation related to mental health problems and suicidality in young people? *Arch. Gen. Psychiatry* **1999**, *56*, 876–880. [CrossRef]

70. Gilman, S.E.; Cochran, S.D.; Mays, V.M.; Hughes, M.; Ostrow, D.; Kessler, R.C. Risk of psychiatric disorders among individuals reporting same-sex sexual partners in the National Comorbidity Survey. *Am. J. Public Health* **2001**, *91*, 933. [CrossRef]
71. Mays, V.M.; Cochran, S.D. Mental health correlates of perceived discrimination among lesbian, gay, and bisexual adults in the United States. *Am. J. Public Health* **2001**, *91*, 1869–1876. [CrossRef]
72. Pillard, R.C. Sexual orientation and mental disorder. *Psychiatr. Ann.* **1988**, *18*, 52–56. [CrossRef]
73. Saghir, M.T.; Robins, E.; Walbran, B.; Gentry, K.A. Homosexuality: III. Psychiatric disorders and disability in the male homosexual. *Am. J. Psychiatry* **1970**, *126*, 1079–1086. [CrossRef] [PubMed]
74. Saghir, M.T.; Robins, E.; Walbran, B.; Gentry, K.A. Homosexuality. IV. Psychiatric disorders and disability in the female homosexual. *Am. J. Psychiatry* **1970**, *127*, 147–154. [CrossRef]
75. Sandfort, T.G.; de Graaf, R.; Bijl, R.V.; Schnabel, P. Same-sex sexual behavior and psychiatric disorders: Findings from the Netherlands Mental Health Survey and Incidence Study (NEMESIS). *Arch. Gen. Psychiatry* **2001**, *58*, 85–91. [CrossRef]
76. Wilcox, H.C.; Arria, A.M.; Caldeira, K.M.; Vincent, K.B.; Pinchevsky, G.M.; O'Grady, K.E. Longitudinal predictors of past-year non-suicidal self-injury and motives among college students. *Psychol. Med.* **2012**, *42*, 717–726. [CrossRef]
77. Barter, C.A.; McCarry, M.; Berridge, D.; Evans, K. *Partner Exploitation and Violence in Teenage Intimate Relationships*; NSPCC/University of Bristol: London, UK, 2009.
78. Eleuteri, S.; Toso, M. How the smartphone apps can improve your sexual wellbeing. *Int. J. Impot. Res.* **2023**, 1–15. [CrossRef] [PubMed]
79. Martín-Castillo, D.; Jiménez-Barbero, J.A.; del Mar Pastor-Bravo, M.; Sánchez-Muñoz, M.; Fernández-Espín, M.E.; García-Arenas, J.J. School victimization in transgender people: A systematic review. *Child. Youth Serv. Rev.* **2020**, *119*, 105480. [CrossRef]
80. Moyano, N.; del Mar Sanchez-Fuentes, M. Homophobic bullying at schools: A systematic review of research, prevalence, school-related predictors and consequences. *Aggress. Violent Behav.* **2020**, *53*, 101441. [CrossRef]
81. Bishop, M.D.; Ioverno, S.; Russell, S.T. Sexual minority youth's mental health and substance use: The roles of victimization, cybervictimization, and non-parental adult support. *Curr. Psychol.* **2023**, *42*, 5075–5087. [CrossRef] [PubMed]
82. Pistella, J.; Ioverno, S.; Rodgers, M.A.; Russell, S.T. The contribution of school safety to weight-related health behaviors for transgender youth. *J. Adolesc.* **2020**, *78*, 33–42. [CrossRef]
83. Andrzejewski, J.; Rasberry, C.N.; Mustanski, B.; Steiner, R.J. Sexual and reproductive health web sites: An analysis of content for sexual and gender minority youth. *Am. J. Health Promot.* **2020**, *34*, 393–401. [CrossRef]
84. Ferfolja, T.; Ullman, J. *Gender and Sexuality Diversity in a Culture of Limitation: Student and Teacher Experiences in Schools*; Routledge: Abingdon, UK, 2020; ISBN 9781138062313.

MDPI AG
Grosspeteranlage 5
4052 Basel
Switzerland
Tel.: +41 61 683 77 34

Children Editorial Office
E-mail: children@mdpi.com
www.mdpi.com/journal/children

www.ingramcontent.com/pod-product-compliance
Lightning Source LLC
Chambersburg PA
CBHW062055220926
51850CB00010B/2772
9783725878277